Instructor's Manual
THE WORLD OF BIOLOGY

Fifth Edition

Solomon & Berg

Jane Aloi
Saddleback College

SAUNDERS COLLEGE PUBLISHING
Harcourt Brace College Publishers
Fort Worth Philadelphia San Diego New York
Orlando Austin San Antonio Toronto
Montreal London Sydney Tokyo

Printed in the United States of America.

Aloi: Instructor's Manual to accompany World of Biology, 5/e by Solomon and Berg

ISBN 0-03-005948-8

567 021 987654321

Table of Contents

PART 5: ANIMAL BIOLOGY

PART 6: PLANT BIOLOGY

PART 7: ECOLOGY

Preface

This book is meant as an aid to both the inexperienced and experienced teacher. This accompanies the 5th edition of The World of Biology by E.P. Solomon and L.R. Berg. This text may be entirely new to you, or because this is a new edition, the new edition will require adjustments this semester. Hopefully, this instructor's manual will help you substantially. The material in this instructor's guide is divided into several sections:

• **Chapter Overview**. This provides a written summary of the chapter material. This may be of use to seasoned instructors to review before lecturing on this chapter.

• **Lecture Outline**. These outlines are also available on disc, and may be particularly useful to new instructors, or instructors who have not taught this course before. The outline closely follows the sequence in the text.

• **Research and Discussion Topics**. These are suggestions which may be used for essay questions, class discussion, or student research papers.

• **Teaching Suggestions**. These are some ideas that I find useful in teaching my Introductory Biology course. Some are "compare and contrast" handouts that my students find useful, such as handouts comparing mitosis and meiosis, photosynthesis and aerobic respiration, and biogeochemical cycles.

• **Suggested Readings**. These are mostly very current articles that are aimed at enriching your lectures. They are not necessarily written at the level of our students, but might be useful for their research papers or other projects. I have tried to include the most recent and interesting articles which will be readily available in your library. I have not included general text references; most of us already have references on chemistry, cellular biology, ecology that we rely on.

The Course Syllabus

One very important component of our courses is the syllabus. This is the introduction to your course, and is an important document that students will refer to during the entire semester. My syllabus has been evolving over the years, with additions and deletions and suggestions from colleagues. I have included a sample syllabus which can be adapted for your course. Not all of the components of my syllabus are applicable to your course, but I hope you will get some good ideas.

A few comments about my syllabus:

• I teach a large lecture section, and make-up exams get "out of hand" at times. To avoid too many make-up requestions, I require that students get in touch with me prior to the exam, or I will not consider a make-up. I also evaluate it on a case-by-case basis. Often students who call in "sick," suddenly get well when you tell them that their make-up exam will consist of essays. However, I believe that it is important to be reasonable. For example, I teach at a college in which many students commute, and sometimes our freeways become a mess! Also, many students are returning to school after many years off, and they are often responsible for their children, whose illnesses may interfere with the student's attendance.

• Cheating and plagiarism is on the increase, and of course is of concern. I make my policies very clear at the beginning, and actually have students sign and turn in a statement indicating that they understand what constitutes cheating. I'm not certain if it would hold up in court, but it sets the ground rules. Cheating in my class has declined significantly since I have adopted this hard line.

• I require students to complete a research paper on a biological subject. Even though I have more than 100 students in this course, grading the papers isn't that big of a burden, and I think it is critical for us to encourage student writing. Because students occasionally purchase research papers from "research" companies, I require them to turn in original copies (these companies sell them Xerox copies). I also tell them not to waste their money, as I can spot one of those purchased papers a mile away. If I suspect plagiarism, I ask the student to visit my office hour, and then I ask them to give me a two minute summary of their research paper.

• I've also included a student information sheet that I ask all students to fill out. It's handy to have their phone numbers in case you need to call them, and I find it useful to know more about them to deal with them on a personal level. For example if a student is a single mom with two kids and a full-time job, I can better understand uneven test scores.

Notes for New Instructors

I have carried out several orientations for new instructors. Often I find that new instructors don't know what questions to ask at the beginning of the semester; then when the questions <u>do</u> come up, there's no one around to answer them. What follows is a list of things you might think about before the semester begins, and find out the answers to them!

• What is the school policy on attendance? Do I have to take attendance?

• Where do I get grade sheets? What information must I fill out on them? Where do I turn in grades at the end of the semester?

• What are my responsibilities for adding and dropping students, if any?

• Do we have tutoring services on campus? Do we have a center for students with special needs? Will I need to make special accommodations for students with special needs on test days?

- How do I duplicate materials? Where is the duplicating center? Do I need signatures on duplicating requests? What is their turn-around time?

- Where is the Scantron machine? How do I make out a key, and run exams?

- What audiovisual materials are available? How do I order them, and the equipment to show them?

XXXXXXX College

Introduction to Biology
Biology XXX

Instructor: XXXXXXX
Spring 1995

• Introduction

Introduction to Biology is designed to provide the fundamentals of biological science. It is fully transferable to the University of XXXXXXXX, XXXXXX State Universities and other four year schools as equivalent to their introductory biology courses.

This course consists of X hours of lecture per week and X hours of laboratory. The units covered are: biochemistry, cellular organization and function, cellular energetics (plants and animals), genetics, reproduction and heredity, the origin and evolution of living things plant and animal diversity, structure and life processes, and ecology.

• Assigned Readings

The textbook for this course is The World of Biology, by Solomon and Berg, 5th edition. Assigned readings are listed in the course schedule. Although the exam questions will be primarily taken from the lecture material, the textbook is a valuable resource for further clarification and learning. Often I find that the difference between an "A" student and a "B" student is that the "A" student utilizes the text and furthers his or her learning by studying it regularly. Please bring your text to class, as well as to lab, as you will find reference to the diagrams in the text of critical importance.

• Course Grading

There will be 4 exams; 3 midterms and a final. The exams will comprise XX% of the lecture component of the course grade. The four exams will contribute XX%, XX%, XX% and XX% respectively to the lecture grade. The final will not be cumulative in nature. Exams will be composed of multiple choice, matching, and true/false questions as well as short answer and essay questions. On exams, you should bring a pencil and GOOD eraser.

Under **no** circumstances will missed exams be made up unless I am notified **prior** to the exam. Missed exams will be made up at my discretion, and may consist of oral and written essays. If the exam is missed due to a medical reason, a doctor's note is required.

The lecture portion of this course contributes XX% to the overall course grade. Grading of exams and the course overall will be based on a straight percentile basis (as opposed to the "curved scale"); >90% = A, 80-89% = B, 70-79% = C, 60-69% = D, <60% = F. Any question concerning any exam question which was incorrectly graded or interpreted will be evaluated. The process for this evaluation consists of the student turning in a written description of the error or discrepancy, along with the exam. This must be done within 2 class periods after the exams are returned, and I will then evaluate the question and make any necessary adjustments. At the end of the semester, all grades will be individually evaluated, and some allowances may be made for lab participation, attendance, etc., if the percentile score is within X% or less of the next grade.

• Research Paper

Each student will complete a research paper on a subject of your choosing (subject to my approval). This paper will contribute X% to the final lecture grade. You must sign up for the topics- no more than two students may write papers on a particular topic. Further, the paper must have some personal significance, or tie-in with your life. This research paper will focus on a biological topic, and will include references to at least X books and articles. The paper will be approximately X-Y pages in length and must be typed. A literature cited section must be included. Papers which do not follow the correct format, or which are late will not be accepted. A paper which is not accepted or not turned in will result in a 5% deduction from your lecture average. Further, Xerox copies will not be accepted. You are encouraged to look at copies of several extremely well prepared papers which I will put on reserve in the library. It is hoped that this paper will illustrate and reflect the ability of the student to critically evaluate and summarize published material.

• Attendance

You are expected to attend all class meetings, or to make up all work for classes missed. Because of the accelerated speed of this course, there is little time to catch up on missed work. Missing more than one class period may result in your becoming hopelessly behind. Further, the majority of the material covered on the exams comes directly from class lectures, so it is to your benefit to attend all classes.

Your registration for this class and this class syllabus represent a contract. If you make the decision to withdraw from this class, you are responsible for meeting the required deadlines and for filing the proper paperwork. I would, however, appreciate a consultation before you withdraw.

• Academic Integrity

According to the official Policies and Procedures of XXXXXXX College "cheating or plagiarism in connection with an academic program" is prohibited and students "may be disciplined... for the following cause." In the rare instance that a student in this course cheats or plagiarizes material, that student will receive a grade of F for this course. Plagiarism is defined as copying from a written source verbatim without putting the material in quotes and citing the source. This word comes from the Latin root *plagiarius,* which means an abductor, or to steal. Included in the realm of cheating is the presentation of work copied from another student as your own, such as copying from another's lab book. I take cheating and plagiarism very seriously!!! I hope this is the last time this subject comes up this semester.

• Tips for success

• Remember that the amount of time spent studying in relation to the amount of time you are in class is recommended to be 4:1. Study time ratios for the lab section will probably turn out to be about 2:1. Therefore, multiply the number of hours you are in lecture and lab each week, and then multiply this by 3 to compute the average number of hours you should be studying per week! You are going to be busy this semester!

• Make copies of the semester schedule. See how it fits in with your other academic, work or personal schedule. Put copies of this schedule on your refrigerator, your bulletin board, and in your notebook.

• Your workspace at home is important. Get a selection of favorite pens, pencils, markers, erasers, Post-It notes, blank 3x5 cards (buy several hundred, as a start), white-out and a stapler. Give yourself some inspiration- tack up a favorite photo, cartoon or quotation.

• Peruse each chapter in the text before really digging into it. Put a post-it note at the end of the chapter so you won't have to waste time constantly seeing "how many more pages" are in the chapter.

• Look at the chapter learning objectives and key concepts in the beginning of each chapter to get a feel for the type of subjects you will be learning.

• Note that there is a chapter summary at the end of each chapter. Make sure that you can define all of the selected key terms.

• To prepare for exams, follow these 5 steps:

1. Start early. You typically need 2 weeks to prepare for a test. Notice that the first lab test in just a few weeks. Start tonight! :)

2. Make and use flash cards. Study them in "down time"- waiting in supermarket lines, while preparing dinner, during commuting (but only if you're not driving).

3. Try to understand concepts, not just memorize facts.

4. In a class such as this, it is optimal to form study groups early! Meet other students in your lecture or laboratory section, exchange phone numbers, and start studying together soon! It will benefit you to begin making flash cards and working on the study guides immediately!

5. Work hard! Success in this course comes to those who are dedicated, persistent, and enjoy the learning process.

Office Hours. XXXXXXX. Office XXXXX phone number: XXX-XXXX. An answering machine is located in my office to take your messages if I am not there.

A Few "Odds and Ends"

• Please do not bring guests, children etc. to class. We cannot allow this due to liability concerns.

• The use of tape recorders is fine; many students like the auditory reinforcement.

• You may need to purchase a dissecting kit from the bookstore. We will not need them for several weeks, but I suggest that you purchase them early, as the bookstore may run out of them. If you have access to instruments and want to compile your own kit, you need scissors, a probe, tweezers and a scalpel at the minimum. You may also elect to purchase some sort of protective clothing for the dissections, or plan to wear older clothes during those labs. You may also want to purchase a box of thin disposable plastic gloves during the dissections.

• You also need to bring a pencil and eraser for the lab drawings. Many students enjoy doing the drawings in colored pencils.

Student Information Sheet

Biology XX
XXX Semester 19XX

____ __________________
Name (preferred first name, last name) student #

____
phonetic pronunciation of name, if necessary

____ __________________
Address Phone #

____ __________________
Employer Hrs Employed per week

____
Position

____ __________________
Prior Colleges attended Date of HS Graduation

____
Previous Science Courses (HS or college)

I am taking this course because____________________________________

My career goals are: ____

I am a ______________________ science student

I anticipate earning a grade of __________ in this course.

If you have any learning disabilities, it would help me to know what these are. You do not have to give me this information, and if you do, it will be kept strictly confidential. ____

Do you have any fears or concerns about science, or this class? _____________
____

Think about your learning style. Do you think you learn best by seeing, by hearing or by doing? Circle one: By Seeing By Hearing By Doing

What is your favorite area of study?_______________________________________

Please tell me more about yourself on the back of this sheet. Include hobbies, family, anything else of interest!

I have read the information in the course syllabus. I am aware of the policy concerning missed exams or other deadlines. I understand that if I do not notify the instructor prior to the exam, <u>and</u> have a valid excuse, an exam may not be made up. Further, I understand the policies concerning plagiarism and other forms of academic dishonesty. I understand that I will receive a grade of F in this course if academic dishonesty occurs.

______________________________ ____________________

name date

Chapter 1. Introducing the World of Biology

Chapter Overview

This first chapter provides an overview of the wide scope covered in the study of biological sciences. Since biology is the study of life, the parameters of life are discussed. Although life itself is difficult to define, living things share certain characteristics. Living things are composed of cells; cells are the fundamental and functional units of life. Cells have various ways of capturing and systematically using energy and materials (metabolism). These metabolic processes are regulated by homeostatic mechanisms.

As Theodosius Dobzhansky once said, "nothing in biology makes sense except in the light of evolution." This will be a unifying theme in this book. All modern organisms have arisen from preexisting organisms by organic evolution. Further, these organisms have evolved by the mechanism of natural selection.

Growth and reproduction are also properties of living systems. Reproduction of cells involves either the formation of new cells for growth, repair and replacement, and the production of a new individual.

Various levels of organization can be identified. Components of living things may be described on a chemical level. Cells are the smallest living unit, and may be arranged in multicellular organisms into tissues, organs and systems. Organisms have a characteristic form and life cycle.

Organisms can also be classified at various levels, which reflect evolutionary ties. The study of this classification is known as taxonomy. The most broad level of classification is the kingdom, and most classification schemes recognize 5 kingdoms of living things.

Living things may also be classified according to their "roles" in the environment; whether they are producers, or varying types of consumers. All organisms interact with their environment. The study of these interactions is known as ecology.

Lecture Outline

<u>Study of Biology</u>

- Current importance of biological studies
- Role of today's biologists

<u>Characteristics of living things</u>

- Cells are the basic living units
- Growth and development characterizes living things

- Metabolic processes also define life
  - Cellular respiration, photosynthesis
  - Homeostatic control of metabolic processes
- Movement is characteristic of living things
  - How plants move- less obvious
  - Animal movement
- Responses to stimuli
  - Simple taxes
  - Complex behaviors
- Reproduction as multicellular growth, or unicellular reproduction
  - Role of DNA in cellular reproduction

<u>Evolution</u>

- Define species
- Define population
- Evolution as a progressive change in living things over time
  - Natural selection as the mechanism of evolution
- Darwin's theory
  - Populations are characterized by heritable variation
  - Some variations are more adaptive and individuals possessing those traits reproduce at a greater rate (differential reproduction)
  - Over time populations change (natural selection)

<u>Various levels of biological organization</u>

- Chemical organization
  - Atoms, molecules
- Cellular organization
  - Organelles
  - The cell is the basic unit of life
- Multicellular organisms
  - Tissues, organs, organ systems
- Ecological organization
  - Communities, ecosystems, the ecosphere
  - The study of ecology
  - Producers, consumers and decomposers
- Taxonomic organization
  - Genera and species
  - Larger classifications: phyla (divisions), class, orders, families
  - 5 kingdom classification scheme
    - Kingdom Prokaryotae (Monera)
    - Kingdom Protista
    - Kingdom Fungi
    - Kingdom Plantae
    - Kingdom Animalia

Research and Discussion Topics

- Justify the current classification of living things into 5 kingdoms. What are alternative views? What evidence do some scientists have to propose a new kingdom?

- How might you justify preserving a certain tract of land to preserve a small, relatively unknown animal. For example, how might you justify your city spending a million dollars to preserve a tract of land which is the only known habitat of a tiny butterfly?

Teaching Suggestions

In my introductory lecture, I discuss what E.O. Wilson has referred to as biophilia, our innate attraction to living things. We have pets, we grow houseplants, we visit zoos and parks and seashores.

I also emphasize the information explosion in the life sciences. In some areas of biological sciences, our knowledge is doubling in less than a decade. Particularly in molecular and cellular biology, our understanding is rapidly expanding, particularly with the innovation of new technological methods. This "information explosion" requires a new generation of scientists. We list some famous scientist on the board, with input from students. Then I point out that 90% of all scientists are alive today. It is important that today's students develop an awareness and knowledge of biological sciences.

Suggested Readings

Raven, P. "Opportunities in Biology." *Bioscience*. May 1990. 385-387. This article describes the current need for a public informed on biological issues.

Saigo, R.H. and B.W. Saigo. "Careers in Biology III". Publication of AIBS and NABT; for copies, write AIBS or Carolina Biological Supply Co. 1988. 4p. An excellent description of careers in biological sciences, with information on opportunities, salaries, traning requirements etc.

Bierzychudek, P. and C.G. Reiness. "Helping nonmajors find out what's so interesting about biology." *Bioscience*. February 1992. 42(2):125-127. A course designed to include elements of the historical, political and social context of biology.

May, R.M. "How many species inhabit the earth?" *Scientific American*. October 1992. 42-48. An account of the search to quantify species on earth, from Aristotle, to some of the more recent, albeit controversial tropical studies.

Benditt, J., ed. "Women in science, 1st annual survey." *Science* special section, 13 March 1992. Nine terrific articles about the increasing role of women in science.

Odum, E. "Great ideas in ecology for the 1990's." *Bioscience*. July/August 1992. 542-545. Odum lists his "top 20" concepts in ecology. Provides some concepts which may be used for class discussion or research paper topics.

Wilson, E.O. "The diversity of life." (Book excerpt). *Discover*. September 1992 46-68. An excerpt from his 1992 book, includes discussion of tropical deforestation.

O'Neill, K.E. "Introduction to classification: kingdoms." *American Biology Teacher*. November/December 1990. 52(8): 495-496. A classroom exercise analyzing organisms and placement in kingdoms.

Holloway, M. "A global view." *Scientific American*. August 1994. 76-83. An excellent description of women's health issues, on a global view. Includes amazing statistics on domestic violence, AIDS, and family-planning.

Chapter 2. The Process of Science

Chapter Overview

Students are introduced to the idea that science is not just a collection of facts. Science is a process, characterized by a classic series of steps. It could also be described as the investigation of rational concepts which are capable of being tested. The principle of biogenesis is used to illustrate the process of science. Early beliefs in spontaneous generation were disproved by Redi and others and replaced with the idea of biogenesis; life comes from life. The first steps in the scientific process include observing natural phenomena and generating hypotheses. Hypotheses are the possible explanations for observed phenomena.

Their experiments included an experimental group and a control. A control is a group which is similar to the experimental group in all respects except for the variable to be tested. In this type of experiment, it is critical that both the experimental group and the control group are treated exactly the same in all ways <u>except</u> for the variable being studied in the experiment. Through these controlled experiments, Redi and others showed that simple organisms do not arise from mud, decaying meat, or culture media.

Scientists use both deductive and inductive reasoning. Deductive reasoning makes predictions about specific circumstances based on general information. Inductive reasoning makes predictions about related facts based on specific experiments. Much of science is based on induction. Experimentation and reasoning may be combined in the <u>scientific method</u>.

Much of science is based on a combination of hard work, good experimentation, and a little luck. Many of the stories of major scientific breakthroughs, such as those made by Pasteur, Fleming are illustrative.

Scientific studies must include not only controls and experimental groups, but the experiment must be sufficiently replicated to gain statistical significance. Also, all types of sampling error and any type of bias must be analyzed.

In science, large unifying concepts are known as theories. In contrast to common usage of the word, scientific theory is not just a hunch or a guess. It is a collection of many experiments and observations which describe a unifying concept. In science, we may discuss the atomic theory, the cell theory or the atomic theory. Over time, theories become so well accepted that they may be referred to as principles.

The process of science also includes ethical considerations. Scientific experiments and their results are published in scientific journals, which are reviewed by other scientists and describe the work done in sufficient detail for others to replicate. Ethical issues facing scientists today include the ethics of research on human issues, as well as genetic manipulation of living things.

Lecture Outline

Science as a process
- Biogenesis as an example
  - Historical perspective
    - Spontaneous generation; nonlife--> life
    - Believed that meat gave life to maggots, mud to tadpoles
  - Francesco Redi (1668)
    - Uncovered meats got maggots, covered meat did not
    - Disproved spontaneous generation
  - Joblot, Spallanzani; experiments on microorganisms
  - Pasteur
- Deductive and inductive reasoning
  - Deductive reasoning
    - Generalities--> specifics
  - Inductive reasoning
    - Specific examples--> generalities/related circumstances
- Scientific method
  - Observations of natural phenomena
  - Describe the problem
  - Hypothesis(es) development
  - Develop experiments to test hypothesis(es)
  - Careful experimentation
  - Analyze results

Hypothesis testing
- Define: hypothesis
- Hypotheses must be testable
- Experimentation
  - Experimental group receives the treatment
  - Control group does not receive the experimental treatment
  - Replication allows statistical analysis
- Sampling errors
- Experimental bias must be considered
  - Bias of the experimental protocol
  - Bias of the experimenter

Results of experimentation may support or modify current understanding
- Define: theory
- Examples of theories in science
- Principles- theories which have been accepted for a long time
- Laws- principles which have exceptional importance, e.g. law of biogenesis

Ethics in science

Research must be published

Peer review

Description of methods and analysis sufficient for repetition

Commitment to truthfulness

Ethical issues in science

Research on humans, other animal experimentation

Research in genetic manipulation

Research and Discussion Topics

• Science as a process, particularly in the past, has been very closely tied to the interests, background and personality of the scientist. Consider the story of Mendel, Darwin and others. The story of the Piltdown man is another. Do research on the principal "players" in the Piltdown hoax and relate the story of the fossil to the story of the people.

• Much of science is based on a combination of hard work, good experimentation, and a little luck. For example Darwin's colleague, Lyell, the geologist said of natural selection: "How obvious! How stupid of me not to have thought of it myself." Most important breakthroughs have been done by well trained scientists, but sometimes being "too close to the problem" creates difficulties. Consider the "race" to discover the structure of DNA. Linus Pauling and his group may have been biased by their work on the single helix of protein, and may not have envisioned DNA as a double helix. Relate this to scientists working on current research projects. Do you think that most of our current breakthroughs are related to serendipity or luck, or just plain hard work?

Teaching Suggestions

• When discussing biogenesis, I point out to students that there is one "exception" to the fact that life comes from life. I ask them to think about it for a moment, then ask for responses. We discuss the idea that the first life on earth indeed came from nonliving components, and I tell them that this will be a subject we will investigate in detail later in the semester.

Suggested Readings

Edey, M.A. and D.C. Johanson. *Blueprints, solving the mystery of evolution*. 1989. Penguin Books. A very readable account of our search to understand the mechanism of evolution, from Linnaeus, to Darwin, to the modern geneticists.

Roman, M.B. "When good scientists turn bad." *Discover*. April 1988. 50-58. Medical fraud and the scientists who uncover fraud.

Culliton, B.J. A series on conduct in science. 5 articles: *Science*. 24 June, 1 July, 29 July, 30 September, 4 November 1988. A lengthy editorial on ethics, error and misconduct in science.

Chapter 3. The Chemistry of Life: Atoms, Molecules and Reactions

Chapter Overview

A discussion of inorganic chemistry begins with a description of elements and their components. The smallest unique component of an element is an atoms, which is itself composed of subatomic particles. In a chemical reaction, the most important (biologically) subatomic particle is the electron, which may be shared (covalent bonds) or gained or lost (ionic compounds), depending on the number of valence electrons. Covalent bonds are particularly important in the bonding between hydrogen, carbon, oxygen and nitrogen atoms, and form the basis for complex organic molecules. Covalent bonds may involve equal sharing of electrons (nonpolar covalent bonds), or unequal sharing of electrons (polar covalent bonds. Ions are characterized by one atom donating a few valence electrons to another atom. Ions dissociate in water (which is an excellent solvent for ionic compounds), and the individual cations and anions become surrounded by water molecules.

Redox reactions are very important in biological systems, as will be seen in future units covering cellular respiration and photosynthesis, for example. Oxidation involves the loss of electrons, reduction involves the gain of electrons; they occur coincidentally and are referred to as redox reactions.

Water is characterized by polar covalent bonds, resulting in the hydrogen atoms being electropositive and the oxygen atom being electronegative. This polar molecule readily forms weak, transient bonds with adjacent water molecules called hydrogen bonds. The hydrogen bonds are critical in allowing the evolution of life on earth as we know it, as it results in a variety of biologically important unique characteristics.

Substances which are polar tend to dissolve in water (are hydrophilic), due to the fact that they form hydrogen bonds with the water molecules. Hydrophobic compounds do not dissolve in water. Because ions dissolve in water, the aqueous medium of the cytoplasm of the cell facilitates chemical reactions.

Because of the hydrogen bonds between adjacent water molecules, water exhibits a great surface tension (only mercury has a greater surface tension), and adhesion and cohesion, which results in capillary action of water in soil. The high thermal stability of water is due to hydrogen bonding and is critical both to aquatic and terrestrial organisms. Finally, because ice floats, life is possible in freshwater habitats in cold environments.

Because life exists in an aqueous medium (the cytosol), understanding of acids (substances which dissociate in water to yield H^+ and an anion), bases (substances which dissociate in water to yield OH^- and a cation) and buffers (substances which resist changes in pH) is critical. When acids and bases are combined in water, salts are formed. Salts, acids and bases dissociate in water, and are known as electrolytes. Electrolytes are the major inorganic compounds in cells.

Lecture Outline

Elements

- Define: substances that cannot be broken down into simpler substances and still maintain their characteristics
- Examples:
  - 92 naturally occuring elements
  - Chemical symbols
    - One or two letters
    - English or Latin names
- Most important elements
- Trace elements

Atoms

- The basic unit of an element
- Size
- Subatomic components and charge
  - Electrons
  - Nuclear components: protons and neutrons
    - Protons
    - Neutrons
- Atomic number: the number of protons
- Atomic mass: the number of protons and neutrons
- Electron shells/ energy levels
- Electron orbitals
  - 1st shell/ orbital can contain only 2 electrons maximum
  - 2nd shell
    - Up to 4 orbitals
    - 8 is the stable number
  - Bohr models are diagrams of the atomic structures

Molecules and compounds

- Molecules are combinations or two or more atoms
  - Examples:
- Compounds are made of molecules of two or more different elements
  - Compounds are characterized by fixed ratios of elements
  - Examples:
- Chemical formulas
  - Element names with subscripts for numbers of atoms
- Structural formulas
  - Two dimensional depictions of molecules
- Chemical equations
  - Chemical formulas of reactant(s) -->
  - Chemical formulas of product(s)
  - Reversible reactions and equilibrium

Chemical bonding

- Reactions involve interactions between outer shell electrons
  - Noble gases are inert; outer shell is filled (8 electrons)
  - Reactive molecules have fewer than 8 valence electrons
- Two primary types of bonding; covalent and ionic bonds
  - Covalent bonds
    - Definition: sharing of a pair of electrons
    - Examples:
      - H_2
      - H_2O
    - Number of valence electrons in biologically important atoms:
      - H- one
      - C- four
      - O- six
      - N- five
    - Single bonds- sharing of one electron
    - Double bonds- sharing of two electrons
    - Triple bonds- sharing of three electron
  - Nonpolar covalent bonds
    - When atoms have equal affinity for electrons (equal electronegativity)
    - Examples: C-H, H-H and O=O
  - Polar covalent bond
    - When atoms have different affinity for electrons (different electronegativities)
    - Examples: H_2O
      - Hydrogen atoms are electropositive
      - Oxygen atoms are electronegative
  - Ionic bonds
    - One atom loses electron(s), another gains electron(s)
    - "An extreme case of polarity"
    - Biological importance: muscle contraction, nerve impulse transmission, other vital life functions
    - Atoms with 1, 2, or 3 valence electrons tend to lose electrons
      - Cations
    - Atoms with 5, 6, or 7 valence electrons tend to gain electrons
      - Anions
    - Ionic compounds= anion+cation
    - Example: Na^+ and Cl^-; sodium chloride
  - Ions in water
    - Dissociation: ions separate into anion and cation in water
    - Water is an excellent medium to dissolve ionic compounds
    - Hydration: ions become surrounded by the polar water molecules
  - The water molecule-H_2O
    - Electronegative part: oxygen atom
    - Electropositive part: hydrogen atoms

Hydrogen bonding- weak bonds between adjacent water molecules
Biological importance: in liquid, solid and gas phases of water
DNA structure

<u>Chemical reactions; oxidation and reduction (redox) reactions</u>

Oxidation is the loss of electrons
Reduction is the gain of electrons
Must happen coincidentally
Biological importance: cellular respiration, photosynthesis

<u>Inorganic compounds</u>

Relatively simple compared to organic compounds
Organic compounds are typically carbon based
Simple inorganic compounds may contain carbon
CO, CO_2, $CaCO_3$
Biologically important inorganic compounds are water, acids, bases and salts

<u>Water</u>

Why is water so important to life on earth?
Makes up the majority of living things
Water as a habitat
Important chemical reactant
Water molecules are polar
Important in the structure of liquid water and ice
How substances dissolve in water:
Hydrophilic substances have an affinity for water; form hydrogen bonds
Hydrophobic substances are insoluble in water (are nonpolar compounds)
Water is a <u>nearly</u> universal solvent; for ions and polar compounds
Water in cells provides a medium for chemical reactions
Cohesion and adhesion is due to hydrogen bonding
Surface tension
Provides a microhabitat for neustonic animals like water striders
Capillary action
Important in providing soil water for plant roots
Water has a very high specific heat
Temperatures of land dwelling organisms (which are <u>made</u> mostly of water remain relatively stable during air temperature changes
Temperatures of bodies of water remain relatively stable during air temperature changes. Important for aquatic organisms
Water has a high heat of vaporization, allowing evaporative cooling
Water is most dense at 4° C, therefore ice floats
Important in allowing organisms to live in lakes during cold winters

Acids, bases and salts

Acids ionize (dissociate) in water and produce H^+ (a proton) and an ion
- Characteristics and examples:
  - Hydrochloric acid- HCl
  - pH below 7

Bases ionize (dissociate) in water and produce OH^- and a cation
- Bases are proton acceptors
- Characteristics and examples:
  - Sodium hydroxide- NaOH
  - pH above 7

pH scale measures acidity or alkalinity
- pH is the negative log of the H^+ concentration
- pH of pure water is 7
- Scale is logarithmic (like Richter scale), so a solution of pH 5 has 100x greater H^+ concentration than a solution of pH 7
- Biological examples
  - Interior of most cells is around 7.3
  - Gastric juices have a very acidic pH
  - Acidic precipitation and deposition
    - Acid "rain" is below pH 5.5
    - Great effects on plants and animals, particularly in the Eastern United States and Europe

Salts
- Salts form from acids and bases
- Example: $HCl + NaOH \rightarrow H_2O + \underline{NaCl}$
  - H^+ from HCl and OH^- from NaOH combine to form H_2O
- Are the main inorganic compounds in a cell- electrolytes
  - Dissociate in water, can conduct an electrical current
  - Other compounds like sugars and alcohols do not- are nonelectrolytes
  - Electrolytes are important in nerve impulse transmission, muscle contraction and many other vital functions

Buffers
- Resist changes in pH; important in homeostasis
- Either accepts of donates hydrogen ions
- Composed of a weak acid and the salt of that acid, or a weak base and its salt
- Biologically important buffers: carbonic acid and bicarbonate ion
  - Important in the circulatory, digestive systems

Research and Discussion Topics

• Investigate the roles of some of the trace elements in living things. Some are well known like iron and iodine, but what is the importance of copper, selenium, vanadium, silicon or chromium? Include in the discussion the possible deleterious effects of too much of a certain trace element.

• Investigate the biological effects of a particular toxin. Why is lead so toxic? In Northern California, buildup of selenium in marshes has been very harmful to birds. How do various poisons like arsenic work?

• Organisms live in a variety of aqueous mediums. Investigate how organisms can maintain an appropriate salt balance while living in freshwater. How do organisms tolerate extremely high salt concentrations, such as seen in the Great Salt Lake?

Teaching Suggestions

• A nifty demonstration of surface tension may be accomplished if you have an overhead projector. Bring in a petri dish filled nearly to the brim with water. Put it on the overhead projector. If your hands are steady, you will be able to put a razor blade on the surface and it will float there. I use a pair of tweezers to place it on the water. On the screen, they'll be able to see the blade floating on the water.

Suggested Readings

Mohnen, V.A. "The challenge of acid rain." *Scientific American*. August 1988. 30-38. Patterns of deposition in the US, causes, atmospheric and watershed processes.

Schwartz, S.E. "Acid deposition: unraveling a regional phenomenon." *Science*. 10 February 1989: 753-762. Sources of sulfur and nitrogen oxides in eastern north America.

Curl, R.F. and R.E. Smalley. "Fullerenes." *Scientific American*. October 1991: 54-63. A description of fullerene (C_{60}), and it's "relatives,'" buckybabies, bunny and fuzzyballs.

Welsch, R.L. "Stand-up chemist." *Natural History*. November 1994. 34-35. A humorous essay including some humorous chemical "formulas."

Chapter 4. The Chemistry of Life: Organic Compounds

Chapter Overview

Organic compounds are the chemicals of life; they are composed primarily of carbon, hydrogen, oxygen and nitrogen atoms arranged around a carbon skeleton (chain or ring). The four major classes of organic molecules are carbohydrates, lipids, proteins and nucleic acids.

Organic compounds are based on a relatively small number of functional groups, which are bonded to a hydrocarbon skeleton. Most of the functional groups are polar, or negatively or positively charged, and therefore are soluble in water. Further, most organic compounds are polymers, chains of simple monomers. These simple building blocks are the basis for many very large, elaborate molecules. The monomers are bonded by condensation reactions. Polymers may be degraded by hydrolysis reactions.

Carbohydrates are important as energy sources, as well as structural components of cells. Simple sugars (monosaccharides) can be directly utilized as fuel for cells, and may be stored as polysaccharides (starch and glycogen) for later hydrolysis and use for energy. Cellulose and chitin are structural carbohydrates, providing structure for plant cell walls and arthropod exoskeletons respectively.

The most abundant lipids are the neutral fats (true fats) which are composed of a glycerol molecule bonded to one, two or three fatty acid chains. Saturated fats (typically animal fats) are characterized by being solid at room temperature, and have straight tails. Unsaturated fats (oils) are liquid at room temperature, and have double bonds in the fatty acid tails. Other important fats include phospholipids, which are important constituents of cellular membranes, and steroids, which includes cholesterol and the steroid male and female sex hormones.

Proteins are very diverse, very large molecules based on monomers of amino acids. There are 20 naturally occurring amino acids. Proteins are arranged in a characteristic way: the order of the amino acids is known as the primary structure. The secondary structure is most commonly the alpha helix, in which the amino acid chain is coiled. The coil is twisted back and bonded to itself in the tertiary structure. If the complete protein consists of more than 1 polypeptide chain, it has quaternary structure. When the primary, secondary and/or tertiary structure of the protein is disrupted, this is referred to a denaturation. Mutations in the genes specifying the amino acid sequence may also result in disruption of the function of the protein.

DNA and RNA are nucleic acids; polymers of nucleotides. DNA is the cell's genetic information, and RNA is involved in protein synthesis. Smaller, related molecules include ATP (adenosine triphosphate) and cAMP.

Lecture Outline

Organic compounds are based on carbon
- Outer valence with 4 electrons; can form up to 4 bonds
  - Carbohydrates
  - Lipids
  - Proteins
  - Nucleic acids
- Hydrocarbons: skeletons of H and C
- May be in chains or rings

Functional groups
- Groups of atoms attached to carbon skeleton
- R denotes the remainder of the molecule
- Simple functional groups:
  - Hydroxyl: R—OH (alcohols)
  - Amino: R—NH_2 (amines)
  - Carboxyl: R—COOH (carboxylic acids)
  - Carbonyl: R—COH (aldehydes)
  - Methyl: R—CH_3 (methane)
- Many are soluble in water since they are polar (have double bonds)
- Other functional groups have positive or negative charges (except methyl)
- Many organic molecules have two or more functional groups

Macromolecules are polymers of monomers
- Monomers are linked by condensation
  - Condensation reactions remove a molecule of water
- Polymers may be degraded by hydrolysis
  - Hydrolysis reactions break bonds with the addition of water

Carbohydrates
- Simple carbohydrate is made of only three elements
  - C : H : O in a 1 : 2 : 1 ratio
- Function: energy source and structural materials
- Monosaccharides
  - Simplest type of carbohydrate; the monomer
  - 6 carbon sugars- hexoses
    - Examples: glucose and fructose (are isomers)
    - Functions: energy source
  - 5 carbon sugars- pentoses
    - Examples: ribose, deoxyribose
    - Functions: part of the DNA and RNA molecules
- Disaccharides
  - Two monosaccharides covalently bonded
    - Sucrose (table sugar)= glucose + fructose
    - Lactose (milk sugar)= glucose + galactose

Maltose (malt sugar; beer)= glucose + glucose

Polysaccharides

+2 monosaccharides covalently bonded; typically a polymer of glucose

Sugar storage forms in plants and animals

Starch

Characteristics: is the form in which most plants store sugars

Glycogen- "animal starch"

Characteristics: more water soluble than plant starch

Is the form in which animals store glucose

Structural materials

Cellulose

Characteristics: insoluble in water, make up most of cell walls in plants

Very abundant- may make up 50% of all carbohydrates in plants

Most animals cannot digest cellulose; have tough bonds

Microorganisms digest cellulose for cows and termites

Chitin

Characteristics: forms the exoskeleton of insects and crustaceans as well as the cell walls of fungi

Composed of monomers of glucosamine

<u>Lipids</u>

Various chemicals which are hydrophobic

Also composed of C, H and O, but less oxygen than in carbohydrates

Energy storage compounds

Also important in insulation, cushioning, cell membranes

Neutral fats (true fats)

Are the most abundant lipid in living things

Only have C, H and O

Building blocks:

Glycerol subunit- head

Fatty acid chains - "tail(s)"

Saturated Fats

Have all possible hydrogen atoms bonded to the carbon skeleton

Are typically animal oils; solid at room temperature

Unsaturated fats

Have some double bonds in the carbon skeleton, hence fewer H atoms

Are typically oils; liquid at room temperature

Two unsaturated fats are "essential"- they must be supplies in the diet

Monoacylglycerols (monoglycerides)

One glycerol with one fatty acid

Diacylglycerols (diglycerides)

One glycerol with 2 fatty acids
Triacylglycerols (Triglycerides)
One glycerol with 3 fatty acids
Phospholipids
Major component of cellular membranes (plasma membrane, and intracellular membranes)
Glycerol with two fatty acids and a phosphate group
Steroids and related substances
Carbon atoms in 4 rings
Examples: cholesterol, bile salts,
Male and female sex hormones, hormones from the adrenal glands

Proteins

Proteins are a very diverse group of molecules
Functions: structure, cellular metabolism, enzymes
20 different naturally occurring amino acids
Amino acids have 4 parts:
Amino group ($—NH_2$)
Carboxyl group ($—COOH$)
Alpha carbon
R group (side chains)
Essential amino acids: those which we must obtain through our diet
Peptide bond
Forms a dipeptide; ultimately a polypeptide chain
Protein structure:
Primary structure
Sequence of amino acids
Sequence specified by genes
Secondary structure
Three dimensional structure
Alpha helix (Linus Pauling): most common form
Gives elasticity to fibrous proteins
(Beta pleated sheet)- another type of secondary structure
Tertiary structure
Secondary structure becomes contorted; disulfide bonds link various areas of the helix
Quaternary structure (not in all)
Seen in proteins made of more than one polypeptide chain
Example: hemoglobin with two alpha and two beta chains
Changes in protein structure
Mutations in genes may result in incorrect amino acid sequences
Example: sickle cell anemia
Denaturation is disruption of the primary, secondary and tertiary structure
Disrupts the correct functioning of the protein
Typically cannot be renatured
Causes: heat, chemicals, varying pH

Nucleic acids

A polymer of nucleotides
Parts:
5-Carbon sugar (ribose or deoxyribose)
Nitrogen containing compound (base)
Phosphate group
Examples:
DNA and RNA
Function: carry the genetic material in the cells of living things
ATP- adenosine triphosphate
Function: energy carrier in cells
cAMP- cyclic adenosine monophosphate
Function: involved in hormone action

Research and Discussion Topics

• Collagen is a very important protein molecule in animals. Discuss the various parts of the body in which collagen is an important structural molecule.

• Discuss the role of cholesterol in the body. Relate the critical role it plays in the cell membrane of animals to the health problems it causes (atherosclerosis)

• Investigate the ratio of saturated to unsaturated fats in common foods. Analyze various oils with respect to polyunsaturated, monounsaturated and saturated fat content. Which oils are the healthiest?

Suggested Readings

Vogel, S. "The shape of proteins." *Discover*. October 1988: 38-43. The 3-D shape of proteins, methods for altering the shape- "protein engineering."

Flannery, M.C. "Thinking chemically about biology." *American Biology Teacher*. September 1990. 52 (6) 379-382. A description of some organic chemicals, includes designer drugs, biopolymers and carcinogens.

Flannery, M.C. "Collagen: complex and crucial." *American Biology Teacher*. Nov/Dec. 1990. 52 (8): 507-510. A description of collagen, diseases, and role in bones.

Duchesne, L.C. and D.W. Larson. "Cellulose and the evolution of plant life." *Bioscience*. April 1989. 39 (4): 238-241. Evolution of cellulose-based cell walls and its functions.

Chapter 5. Cell Structure and Function

Chapter Overview

The cell theory, which was outlined by Schleiden, Schwann and Virchow in the 1800's states that the cell is the fundamental unit of life, all cells are composed of one or more cells, all cells arise from preexisting cells. Although diverse in structure and function, cells possess a plasma membrane, which encloses the cytoplasm, the aqueous medium which contains various particles and membranous structures.

Prokaryotic cells all belong to Kingdom Prokaryotae, and lack typical membrane bound organelles. Eukaryotes do have specialized membrane bound organelles and includes the protists, plants, fungi and animals. These membranous organelles allow compartmentalization and channelization within the cell. Also, membranes are the site of many metabolic processes, such as part of the processes of photosynthesis and cellular respiration. Cells are necessarily small to maximize the surface area to volume ratio, although their size and shape vary with their functions.

The early study of cells was begun in the 1600's with crude light microscopes, but was greatly aided in the 1950's with the development of the electron microscope. More recent techniques such as cell fractionation have allowed even more detailed studies.

The nucleus houses the genetic material; either in the form of chromatin, the dispersed form seen when cells are not dividing, or in the condensed form of chromosomes, seen during cell division. The nucleus also houses the nucleolus, the site of ribosome assembly. The nuclear membrane is characterized by numerous pores, and is continuous with the endoplasmic reticulum. The rough endoplasmic reticulum is studded with ribosomes, and is the site of protein manufacture. Smooth endoplasmic reticulum is the site of phospholipid and steroid manufacture, and detoxification of various substances. The proteins made by the rough endoplasmic reticulum are passed to the Golgi complex for further modification and sorting before passage to their final destination; either the plasma membrane, or storage within the cytoplasm. Further, the Golgi complex produces vesicles called lysosomes. Lysosomes contain digestive enzymes which "recycle" worn out cellular organelles, as well as particles that the cell has ingested.

Vacuoles perform a variety of functions in plant and animal cells. In plant cells, they store water, ions, and sometimes pigments and toxins. In animal cells, they may act in excretion of water (contractile vacuoles), or as food vacuoles. Microbodies are other membranous organelles of various functions; peroxisomes degrade hydrogen peroxide and glyoxysomes are important in plant seed germination. Very unusual, double membrane bound organelles, which also contain their own DNA are the mitochondria and chloroplasts. Mitochondria function in production of ATP from nutritive molecules, and chloroplasts function

in photosynthesis in plants.
The components of the cytoskeleton give a cell structure and may aid in their movement. Microtubules are large rod-like fibers which are important in chromosomal movement during cell division, intermediate fibers give structure to the cell, and microtubules are small fibers made of actin and are important in cytoplasmic streaming and the formation of pseudopodia. Centrioles are paired structures important in mitosis. Microtubules also are the supporting structures of cilia and flagella. The structure of both of these cellular projections is identical, with a 9 + 2 arrangement of microtubules; the term flagella is used when the projections are long and few in number, the term cilia is used when the projections are short and numerous. Also lending support to cells of plants, bacteria, fungi and some protists are cell walls. Cell walls of plants are strengthened with the polysaccharide cellulose. Other materials such as proteins, pectin and lignin may add further strength.

Lecture Outline

The cell
- Cell theory (Schleiden, Schwann, Virchow)
    - All organisms are composed of one or more cells
    - The cell is the basic living unit of organization for all organisms
    - All cells arise from preexisting cells
    - Cells contain all of the hereditary information
- Three basic structures of all cell types
    - Plasma membrane
    - Organelles
    - Cytoplasm
- Prokaryotes v. eukaryotes
    - Prokaryotes: bacteria and cyanobacteria
        - All are members of kingdom Prokaryotae (old name: Monera)
        - Characteristics: lack most membrane bound organelles
            - Do have DNA, in a concentrated region
            - Tend to grow rapidly, divide often
    - Eukaryotes: protists, fungus, plants and animals
        - Characteristics: possession of specialized membrane-bound organelles
        - Nuclear membrane, and other specialized membranes
- Cell size
    - Range: prokaryotes 0.2-5 μm, eukaryotes 10-100 μm
    - Surface to volume ratio limits cell size
        - Volume increases with the cube of the diameter
        - Surface area increases with the square of the diameter
        - Significance: as a cell grows, volume (cytoplasm) increases at a faster rate than surface area (plasma membrane)

Cell shape is related to function
- Epithelial cells are plate like, roughly square or spherical
- Nerve cells and muscle cells (fibers) are elongate

Cytology: the study of cells
- Anton Van Leeuwenhoek developed one of first microscopes
- Robert Hooke described first cells (dead cells of cork)
- Light microscopes: 1600's to today
- Electron microscopes: 1950's
  - Allowed study of ultrastructure
- Features of a microscope
  - Magnification
  - Resolving power
- Cell fractionation: used to purify cellular organelles for further study

<u>Cellular structures of eukaryotes</u>

Cytoplasm: material outside of nucleus, consists of fluid and particles and membranes

Nucleoplasm: material within the nucleus

<u>Nucleus</u>

Largest organelle

Nuclear envelope
- Double membrane
- Has many nuclear pores

Nucleolus (pl: nucleoli)
- Site of ribosome assembly
- Ultimately ribosomes leave through the pores and are found free in the cytoplasm or associated with the rough ER

The genetic material: DNA
- Chromatin
  - Seen in cells which are not dividing
  - Decondensed DNA associated with RNA and proteins
- Chromosomes
  - Seen in cells which are dividing
  - Condensed DNA in a highly organized and compact form

<u>Endoplasmic reticulum (ER)</u>

Sets of membranes continuous with the nuclear and plasma membranes

Membranes act to divide up the cytoplasm into compartments and channels

Two types:
- Rough ER: have ribosomes
  - Ribosomes- RNA-protein structures
    - May be free <u>or</u> attached to the rough ER

Ribosomal function: development of the primary structure of proteins

Smooth ER: lack ribosomes

Function: phospholipid, steroid synthesis

Golgi complex

Stacks of flattened membranes (look like a stack of pita bread)

Sets of smooth membranes derived from the ER

Functions: sorting, modifying proteins

Ultimately transports products to the plasma membrane, or are stored within the cytoplasm

Also produces the lysosomes

Lysosomes

Contain powerful digestive enzymes

Function as a recycling center of the cell

Digests worn-out organelles, or materials the cell has ingested

Rheumatoid arthritis is due to damage in joints due to "leaky" white blood cell lysosomes

Vacuoles

In plants

Fluid filled sacs, similar function to lysosomes

Storage sites for water, ions, toxins, pigments

In animals, includes contractile vacuoles, food vacuoles

Microbodies

Various organelles which regulate different metabolic reactions

Examples:

Peroxisomes break down hydrogen peroxide

Glyoxysomes are important in germinating seeds

Organelles involved with energy production and utilization

Mitochondria- the "power plants" of the cell

Site of cellular respiration, which converts organic molecules to ATP

Double membrane bound; specialized membranes for energy production

Have a small amount of their own molecules of DNA, ribosomes

Chloroplasts

Site of photosynthesis

Also double membrane bound

Inner structure

Stacked membrane system (thylakoids) (3d system of membranes)

Pigments: chlorophyll and others

Also have their own molecules of DNA

Other plastids
- Leukoplasts store starch, are colorless
- Chromoplasts store colored pigments

Structural elements of the cell

"Cytoskeleton"

MTOC (=centrosome), the microtubule organizing center
- Directs construction of cytoskeleton

Microtubules
- Largest in size
- Composed of tubulin
- Radiate from centrosome
- Important in cell division, aid in chromosomal movement

Microfilaments
- Smallest in size
- Composed of actin
- Important in movement of organelles, as well as pseudopodia

Intermediate fibers
- Intermediate in size
- Important in cellular structural support, help maintain cell shape

Centrioles (2)
- Primarily in animals
- Function in cellular division

Cilia and flagella of eukaryotes
- Flagella are few in number and relatively long
  - Flagella function in movement of the cell, such as on sperm
- Cilia are more numerous and relatively short
  - Cilia function in movement of the cell, or in movement of materials over the surface of the cell, such as in the respiratory system or reproductive system
- Structurally are the same, composed of microtubules arranged in a 9 + 2 arrangement
- At base is basal body, which has a 9 + 3 arrangement of microtubules similar to the structure of a centriole

Cell walls

Found in bacteria, protists, plants, and fungi

Function: protection of the cell, adds strength to the organism

Components:
- Major component is cellulose
- Other materials are polysaccharides, proteins, pectins, lignin

Research and Discussion Topics

• Investigate the evidence supporting the endosymbiotic theory, which indicates that cellular organelles such as chloroplasts and mitochondria may have arisen from prokaryotes which may have symbiotically "invaded" other cells.

• Research medical conditions such as Tay Sachs or gout which are caused by malfunctioning cellular organelles. What causes these conditions, and how are they treated?

• Describe the action of the lysosome in various "unique" situations, such as the division of the individual cells during embryonic development, or the part it plays in the endometrium of the uterus during the menstrual cycle, or its role in reabsorption of the tadpole's tail during metamorphosis.

Teaching Suggestions

• I distribute this worksheet for the students to complete, which compares the cellular structure of bacteria, plants and animals.

	Bacteria	Plants	Animals
Prokaryotic or Eukaryotic?			
Nucleus?			
DNA location:			
Chloroplasts?			
Endomembranes?			
Ribosomes?			
RER, SER?			
Cell membranes?			
Cell walls?			

Suggested Readings

Keister, E. " A bug in the system." *Discover*. February 1991. 70-76. A discussion of the function of the mitochondrion, and how a few genetic defects in the mitochondrial genes leads to human diseases.

Story, R.D. "Textbook errors and misconceptions in biology: Cell structure." *American Biology Teacher*. August 1990. 52 (4): 213-217.

Verner, K. and G. Schatz. "Protein translocation across membranes." *Science*. September 1988. 1307-1313. How proteins are translocated across membranes of eukaryotic cells.

McLaughlin, E., J. Giannini and K. Fishbeck. "Color-coded organelles." *American Biology Teacher*. October 1994. 56 (7): 420-423. A lab illustrating diffusion, osmosis and active transport in beet vacuoles.

Chapter 6. Biological Membranes

Chapter Overview

The plasma membrane forms a physical barrier around cells. However, it is much more than a simple barrier; it functions as selective barrier, as well as being the site of various proteins and carbohydrates which function as gates, and recognition and communication sites.

The plasma membrane is now known to be a bilayer of phospholipids, arranged with the fatty acid chains pointing inward and overlapping, and the glycerol and phosphate portion facing the outer boundary of the membrane. These molecules have been experimentally shown to be very fluid; to move laterally, and even sometimes switching "sides." This amphipathic molecule acts to prevent the movement of many types of molecules through the bilayer. Integral proteins which are also amphipathic may be embedded in the membrane, and may act as channels for certain molecules to pass. The peripheral proteins are found on either side of the bilayer, typically bonded to the end of the integral proteins. Peripheral proteins often have short carbohydrate chains attached to them, and act as recognition or communication sites. Projections of microfilaments may cause the plasma membrane to have many foldings, known as microvilli. These increase the surface of the cell and are particularly prominent in intestinal cells.

Cells may be joined in a variety of ways. Plant cells have plasmodesmata, and animal cells have similar openings between adjacent cells known as gap junctions, particularly common in heart muscle cells. Other tough "rivets" between cells, desmosomes, connect cells subject to considerable stretching. The abundant tight junctions hold cells tightly together, very important in epithelial tissues.

Materials may pass in and out of cells in a variety of passive or active mechanisms. Diffusion is a passive mechanism in which molecules move from areas of high to low concentration. Small molecules such as oxygen and carbon dioxide, and ions may diffuse into cells. Diffusion is most rapid at elevated temperatures, and with small molecules. Hydrophilic molecules like glucose and amino acids move into a cell via special channels by the process of facilitated diffusion.

Osmosis is the process of diffusion of water molecules from a region of higher to lower concentration. When a cell is placed in a hypertonic solution, it will tend to lose water, as water is more highly concentrated within the cell. When a cell is placed in a hypotonic solution, water flows into the cell. In freshwater protists, a contractile vacuole expels this excess water. In plants, the influx of water fills the central vacuole, and the resulting water pressure gives the plant turgor pressure, important in keeping herbaceous plants erect.

Active transport mechanisms require much energy, in the form of ATP. Active transport mechanisms allow a cell to move molecules <u>against</u> the concentration gradient. Perhaps the most important active transport mechanism is the sodium-potassium pump, which actively moves sodium out of cells and potassium in, via specialized membrane proteins. Movement of large quantities of material also requires energy, and may be described as exocytosis, in which cells expel the contents of cellular vesicles, and endocytosis in which cells envelop materials from the environment, including phagocytosis, pinocytosis and receptor-mediated endocytosis.

Lecture Outline

<u>Functions of biological membranes</u>

- The plasma membrane
  - Is a selective barrier, allowing certain materials to pass in and out of the cell
  - Allows the cell to receive and transmit information, due to proteins embedded in the membrane
  - Communication with adjacent cells, again via proteins
- Internal membranes
  - Compartmentalization of organelles
  - Sites for enzymes of cellular respiration and photosynthesis

<u>Phospholipid bilayer</u>

- Fluid Mosaic Model
  - 1972, Singer and Nicolson
  - Fluidity: molecules move around a lot
  - Mosaic of proteins embedded in phospholipids
- Size: less than 10 nm thick
- Structure:
  - Fatty acid chain is hydrophobic
  - Glycerol and P group are hydrophilic
  - The molecule is amphipathic
- Spontaneously forms two layers; hydrophobic tails to the inside
  - No bonds hold the molecules; hence fluidity
- Impermeable to ions and polar molecules
  - Water, lipids and nonpolar molecules may pass through

<u>Proteins of the cell membrane</u>

- Functions:
  - Transportation of molecules in and out of the cell
  - Cell to cell recognition, message transmission
- Integral proteins
  - Structure: amphipathic with one region in the hydrophobic layer
    - Firmly embedded in the membrane
- Peripheral proteins
  - Structure: bound to emergent end of integral proteins
    - Can be removed without damaging cell membrane integrity
    - Are mostly glycoproteins

Function: primarily communication and recognition
Glycocalyx= glycoprotein coat of a cell
Functions to glue cells together
Microvilli
Protein microfilaments (actin) produce numerous extensions of the plasma membrane
Most cells have microvilli; are marked in intestinal cells
Greatly increases the surface area of the cell

Cell to cell junctions
Plant cells
Plasmodesmata are openings in the cell wall through which strands of plasma membrane extend
Passageway for water, ions and other materials
Animal cells
Gap junctions are open passageways between cells
Allows electrical communication between cells, as in heart muscle
Desmosomes are protein-strengthened rivets between cells subject to mechanical stresses
Tight junctions hold cells closely together, particularly important in epithelia, preventing leakage

Transmembrane gradients– how molecules move in and out
Plasma membrane is selectively permeable
Diffusion
Define: movement of particles from areas of higher to lower concentration
Materials diffuse down a concentration gradient
Rate of diffusion depends on:
Size: large molecules move less rapidly, diffuse more slowly than small
Heat speeds the rate of diffusion
What does diffuse through the plasma membrane?
Dissolved gases, water, small ions and molecules
What does not diffuse through plasma membrane?
Hydrophilic, or large molecules
Facilitated diffusion
Carrier proteins accelerate passage of hydrophilic molecules like glucose and amino acids through the plasma membrane
Move from high to low concentration, hence "diffusion"
Carrier proteins are specific, and controllable
Osmosis
Definition: diffusion of water molecules across a semipermeable membrane from region of higher to lower concentration of water molecules

Hypertonic: a solution with a greater solute concentration than the cell

Hypotonic: a solution with a lower solute concentration than the cell

Isotonic: a solution with an equal solute concentration as the cell

Examples:

- Distilled water (100% water) is hypotonic to animal cells
- Ocean water is hypertonic to animal cells

Dynamics of osmosis:

- When a cell is placed in a hypotonic solution, it gains water
- When a cell is placed in a hypertonic, it loses water
- When a cell is placed in an isotonic solution, it gains and loses water molecules at equal rates

Animal cells in hypotonic solutions (lakes)

- Contractile vacuoles expel excess water

Plant cells in hypotonic solutions (soil water)

- Central vacuole has great turgor pressure, helps keep plant erect

Active transport

- Allows cells to gain materials in short supply
- Molecules are moved from regions of low concentration to higher
- Requires energy in the form of ATP
- Protein channels (integral proteins) in the plasma membrane accomplish active transport
- Sodium-potassium pump
  - Maintains a high internal K concentration, lower Na concentration

<u>Movement of large particles across the membrane</u>

Exocytosis

- Cells eject wastes or cellular products like hormones or mucous
- Cellular vesicle fuses with plasma membrane, expels product
- Also aids in growth of the cell by adding to the plasma membrane

Endocytosis

- Phagocytosis
  - Cell envelops particles to be ingested, plasma membrane forms a vesicle around it
  - Typically fuses with lysosomes for digestion
  - Bacterial defense by body- white blood cells engulf foreign material
- Pinocytosis
  - Cell takes in dissolved materials with the aid of the microvilli
- Receptor mediated endocytosis
  - Specialized receptors in the plasma membrane bind with material to be enveloped, take it in by endocytosis
  - Is the mechanism by which cells take up cholesterol

Research and Discussion Topics

- Cystic fibrosis is due to a defect in a membrane pump of the cell membrane. Discuss the causes and treatment of this disease.

- Discuss the mechanism by which cells use receptor mediated endocytosis to take in cholesterol. Investigate the connection to the genetic disorder, hypercholesterolemia.

- Compare and contrast the blood-brain barrier to the blood-testes barrier. How are they similar? How are they different? What are the clinical applications of these two barriers?

- Discuss the importance of the cholesterol component in the cell membrane of animal cells. Where is it manufactured? What is its function?

Teaching Suggestions

- Give students examples of diffusion that they can relate to; the diffusion of perfume from the person sitting next to them (or perhaps less pleasant odors!). Describe a real-life experience such as coming home, and smelling the odorous molecules from dinner which is in the oven. Relate both of these examples of diffusion to the fact that increasing temperature increases the rate of diffusion.

Suggested Readings

Leinhard, G.E., J.W. Slot, D.E. James and M.M. Mueckler. "How cells absorb glucose." *Scientific American*. January 1992. 86-91. Glucose transport system, how insulin regulates it.

Vogel, S. "Dealing honestly with diffusion." *American Biology Teacher*. October 1994. 56 (7): 405-107. Interesting facts about diffusion, hints on teaching the subject.

Stossel, T. P. "The machinery of cell crawling." *Scientific American*. September 1994 54-63. How cells move, including the action of actin and myosin.

Various Authors. "Cystic Fibrosis Cloning and Genetics." *Science* Special issue, 8 September 1989. Identification and cloning of the cystic fibrosis gene.

Roberts, L. "To test or not to test." *Science*. 5 January 1990. 17-19 Due to the high frequency of the CF gene, a discussion over whether to start a genetic screening program.

Chapter 7. The Energy of Life

Chapter Overview

Most biological processes are energy requiring processes. Energy flows in a one-way direction through biological systems. Energy may exist in various forms: chemical, nuclear, solar, thermal, electrical and mechanical. Mechanical energy may be stored (potential) or kinetic energy. Energy may be transformed from one form to another, typically with a loss of heat.

The first law of thermodynamics states that energy may not be created or destroyed, but rather may only be transformed from one form to another. The second law of thermodynamics states that as energy transformations occur, entropy increases as heat is given off. As the transformations are not 100% efficient, the amount of usable energy decreases with sequential transformations.

Chemical reactions may be exergonic (heat producing or exothermic) or endergonic (energy requiring or endothermic). Exergonic reactions occur spontaneously, although the two types of reactions are often coupled in biological systems.

ATP is the molecule in which chemical energy is stored in cells. ATP is a nucleotide made of adenine, ribose and three phosphate molecules, in which the energy is stored. One phosphate molecule is easily hydrolyzed, releasing energy in this exergonic reaction. ATP is reformed during cellular respiration in an endergonic reaction. ATP is constantly recycled in cells.

Enzymes are organic catalysts, and are primarily proteins. Enzymes speed up the rate of reactions, although they do not themselves change the outcome of the reaction. They are specific to the reaction, and may be reused many times with no permanent conformational change. Enzymes work by decreasing the energy of activation of a chemical reaction, and act to cleave one substrate molecule into two or more products, or to join two or more substrates forming one product molecule. The induced fit model describes the bonding of the substrate to the active site of the enzyme, where a tight bonding causes a conformational change in the enzyme to fit the substrate more closely. Then the product(s) fall from the enzyme, leaving it ready to react again.

Enzymes often require inorganic cofactors, such as metal ions, or organic nonprotein coenzymes. Further, enzymes often work in sequence, and often the end product acts as an inhibitor of the first enzyme, in feedback inhibition. Other modes of regulation of enzymatic action include genetic controls of enzyme production. Allosteric enzymes are controlled by their allosteric site. When a certain activator or inhibitor molecule binds to the allosteric site, it causes the active site to change shape so either it will or will not bind to the substrate.

Because most enzymes are proteins, their structure is affected by temperature and pH. In the human body, most enzymes work optimally at normal body temperature. Depending on the part of the body, the optimal pH varies. Enzymes may also be affected by chemicals, reversibly or irreversibly. Reversible inhibitors include competitive inhibitors, and noncompetitive inhibitors. Irreversible inhibitors include some insecticides and drugs.

Lecture Outline

<u>Energy in biological systems</u>

- Utilized in nearly all biological activities
- Life depends on energy transfers and transformations
- Energy is defined as the ability to do work
    - Chemical energy is stored in the bonds of molecules
    - Nuclear energy is stored within the nucleus of an atom
    - Solar energy is transported to the earth from the sun
    - Heat energy is thermal energy
    - Electrical energy is the flow of charged particles
    - Mechanical energy is the movement of bodies
        - Mechanical energy may be stored as potential energy or in the form of kinetic energy (energy in motion)
- Energy may be measured in the form of heat
    - The standard measurement of heat is the joule
        - A calorie is the amount of heat required to raise the temperature of one gram of water 1° C

<u>Laws of thermodynamics</u>

- First law of thermodynamics (law of conservation of energy)
    - Energy cannot be created or destroyed, but may only be transformed from one form to another
- Second law of thermodynamics (law of entropy)
    - Entropy (disorder) continually increases, as some energy is constantly "lost" as heat during energy transformations
    - Energy transformations are therefore less than 100% efficient in terms of usable energy

<u>Metabolic reactions</u>

- Chemical reactions involve changes in the molecular structure of substances
- Exergonic reactions release free energy (heat = exothermic)
- Endergonic reactions require energy inputs to proceed (endothermic)
- Chemical reactions are often reversible
    - Reversible reactions reach equilibrium
- Endergonic and exergonic reactions are often coupled in biological systems

ATP

- Adenosine triphosphate stores energy in cells
- ATP consists of:
  - Adenine, a nitrogenous base
  - Ribose, a pentose
  - Three phosphate molecules
    - Energy is stored in the phosphate bonds
    - Are easily hydrolyzed forming ADP
      - This is an exergonic reaction, releasing energy
    - The removed phosphate is typically transferred to another molecule, known as phosphorylation
  - ATP is rapidly recycled in cells

Enzymes

- Biological catalysts
- Nearly all are proteins
- Speed up chemical reactions
  - Affect only the rate of a reaction
  - Are very specific and typically work on one substrate
  - Most are named for the substrate plus the suffix -ase
- Enzymes work by lowering the activation energy
  - May be reused very rapidly
  - Are not consumed or permanently changed in the reaction
- Typical enzymes are proteins, and thus have a 3-D structure
  - The catalytic or active site is the area where the substrate fits
- The induced fit model of enzymatic action
  - Substrate(s) bind to the active site(s)
  - The binding is very tight and the molecules change shape slightly
  - The product(s) formed then fall from the active site
- Enzymes may act to bond two or more molecules together, or break one molecule into two or more components
- Enzymes often require cofactors
  - Inorganic cofactors are typically trace minerals
  - Coenzymes are non-protein organic cofactors, such as vitamins
- Enzymes often work in a series

Regulation of enzymatic action

- Genetic controls act by varying production of enzymes
- Sequential enzymatic processes may be controlled by formation of the product
  - Feedback inhibition
- Allosteric enzymes are controlled by conformational changes which result from binding to the substrate
  - The control may be inhibitory or stimulatory

Effects of temperature on enzyme activity
Temperature affects enzymes because most enzymes are proteins
High temperatures denature enzymes
pH affects enzyme activity
Enzymes in the human stomach work optimally in acidic conditions
Enzymes in the small intestine work optimally in basic conditions
Chemical inhibitors prevent enzymatic action
Reversible inhibitors may act by competitive inhibition
Inhibitors block the active site by binding to it
Reversible noncompetitive inhibition acts by binding at a site other than the active site, but act by changing the shape of the enzyme
Many of these are important in feedback inhibition
Irreversible inhibitors permanently destroy the enzyme
Many insecticides act in this way
Penicillin inhibits a bacterial enzyme involved in cell wall formation

Research and Discussion Topics

• Insecticides such as malathion and parathion act by inhibiting enzymes. Investigate their mode of action.

• Many household items use enzymes. Discuss the action of proteolytic enzymes in laundry detergents and stain removers, as well as meat tenderizers.

• Most animals digest their food by secreting digestive enzymes into a digestive tub (intestine). Several animals digest, or at least predigest their food outside of their bodies. Look up and describe the feeding mechanism of spiders. Starfish feed on bivalves like mussels by secreting digestive enzymes into the mussel to kill and "predigest" it. Describe the mechanism.

Chapter 8. Energy-Releasing Pathways

Chapter Overview

All living things must extract energy from nutritive molecules. We gain our nutritive molecules from the food we eat; plants make their own food molecules, but then must go through the same process to get useful energy from those food molecules. These are known as catabolic reactions, which are exergonic, releasing energy, which cells harness in the molecules of ATP. There are three different types of energy releasing pathways, which are utilized depending on the type of organism, and whether oxygen is present or absent.

Energy releasing pathways couple oxidation and reduction in redox reactions; in this case glucose, or some other molecule is oxidized. The electrons are transferred to electron carriers, NAD^+ and FADH.

Most plants and animals utilize aerobic respiration to produce ATP. There are four principle steps: glycolysis, formation of acetyl coenzyme A, citric acid cycle, and electron transport and chemiosmosis.

In glycolysis, glucose is phosphorylated by the addition of phosphate molecules from ATP molecules. 2 ATP molecules are needed to "jump start" this reaction. The phosphorylated glucose is then changed into PGAL. Then PGAL is modified to form pyruvic acid, producing 4 molecules of ATP in the process, as well as 2 NADH molecules.

Pyruvic acid then enters the matrix of the mitochondrion and is ultimately converted to acetyl CoA. In this series of reactions, 2 carbon dioxide molecules are produced, as well as 2 NADH molecules. In the citric acid cycle, the acetyl CoA molecules are processed, giving off 4 carbon dioxide molecules, and ultimately regenerating the carbon compound (oxaloacetic acid) which starts the cycle again. In the two "turns" of the citric acid cycle, 6 NADH and 2 $FADH_2$ molecules (all electron carriers) are produced. At the end of the citric acid cycle, all of the carbon atoms that were originally present in the molecule of glucose are now in the form of carbon dioxide.

The electron carriers (NADH and $FADH_2$) from the previous steps now pass electrons on to a series of proteins embedded in the inner mitochondrial membranes, the electron transport system. As the electrons are transferred from protein to protein, they lose some energy. Ultimately, at the end of the electron transport chain, the lower energy electrons combine with molecular oxygen and hydrogen ions to form water.

As the electrons are passed to the electron carriers, hydrogen ions are expelled into the mitochondrial matrix. Because of the concentration gradient, hydrogen ions diffuse back to the intermembrane space through specialized protein channels, ATP synthetase. This process is known as chemiosmosis. As they pass through these channels, ADP is combined with inorganic phosphate to form ATP. This is the site of production of the majority of ATP in this process (32 ATP molecules). Compare this to the fact that only 2 ATP molecules were produced in glycolysis and 2 through the citric acid cycle. This totals 36 molecules of ATP per one glucose molecule, although research shows that the actual yield may be less than this.

Nutrients other than glucose may be used to produce ATP. Fats and amino acids may be broken down into simpler molecules which may be shuttled into glycolysis or the citric acid cycle, with varying amounts of ATP produced.

Anaerobic pathways are used in simpler organisms (primarily bacteria), or when oxygen is not present. Anaerobic respiratory pathways are used by bacteria, with the net yield of ATP being only the 2 ATP molecules from glycolysis. Typically, nitrate or sulfate are the final electron acceptor. In fermentation pathways, the final electron acceptor is an organic molecule. In alcoholic fermentation, seen in yeast, the final product is ethyl alcohol and carbon dioxide. This is commercially used in production of baked goods, beer and wine. Lactic acid fermentation is carried out by some bacteria and fungi. In this type of respiration, lactic acid is produced. This is seen in our muscle cells, when oxygen supplies are inadequate. We commercially use bacteria to make cheese and yogurt.

Lecture Outline

Energy-releasing pathways

- Digestion breaks down polymers of nutritive molecules into monomers
    - Catabolic reactions are exergonic, releasing free energy
- Three different catabolic pathways to gain energy
    - Anaerobic respiration
    - Fermentation pathways (also anaerobic)
    - Aerobic respiration (seen in most animals and plants)
        - Long series of reactions, each step catalyzed by a specific enzyme
        - Overall reaction: $C_6H_{12}O_6 + 6\ O_2 + 6\ H_2O$--> $6\ CO_2 + 12\ H_2O$ + Energy
- Redox reaction
    - In reduction reactions, substances gain electrons
        - Oxygen participates in the reduction reaction
    - In oxidation reactions, substances lose electrons
        - Glucose participates in the oxidation reaction
    - Electrons are transferred to electron acceptors, which are then reduced
        - NAD^+ = nicotinamide adenine nucleotide
        - FAD = flavin adenine dinucleotide
    - Along with the redox reaction, chemical energy is released, and used to form ATP from ADP and P_i

Aerobic respiration

(Kreb's cycle)

Four stages: glycolysis, formation of acetyl coenzyme A, citric acid cycle, and electron transport and chemiosmosis

Glycolysis

- Term means sugar splitting
- 1 Glucose molecule --> 2 3-carbon pyruvic acid molecules
- Occurs in the cytoplasm, in aerobic or anaerobic conditions
- 2 phases:
  - First steps require input of energy and ATP (2/molecule of glucose)
    - Glucose is phosphorylated
    - Phosphorylated glucose molecule is split into 2 PGAL molecules
  - Second steps produce ATP (4 molecules)
    - PGAL is oxidized by removal of 2 hydrogen atoms and ultimately transformed into pyruvic acid
    - Hydrogens combine with NAD^+, to form NADH
    - Net gain is 2 molecules of ATP per molecule of glucose

Pyruvic acid is changed into acetyl CoA

- Occurs in the mitochondrial matrix
- The two molecules of pyruvic acid enter a series of steps to produce acetyl CoA
- Accounting for the carbon atoms:
  - Acetyl CoA has two carbon atoms per molecule
  - Carbon dioxide is released (2 molecules total)
- The carbon fragment is oxidized ultimately forming acetyl CoA
  - 2 Hydrogens are picked up by NAD^+, to form 2 NADH

Citric acid cycle (also known as Krebs cycle)

- Final oxidation of all types of fuel molecules
- Occurs in the matrix of the mitochondrion
- 8 steps, each catalyzed by a different enzyme
  - Acetyl CoA combines with oxaloacetic acid forming citric acid
  - Citric acid is processed, losing carbon atoms as CO_2
  - Oxaloacetate is reformed, entering the cycle again
- Two acetyl CoA molecules are formed; must cycle twice to process both
- 2 ATP molecules are produced
- Electron carriers formed: 6 NADH and 2 $FADH_2$

Electron transport system

- NADH and $FADH_2$ pass their electrons to electron receptors
  - Receptors are embedded in the inner mitochondrial membrane
- NADH transfers electrons to the first acceptor
  - $FADH_2$ transfers electrons a little later, with a slightly lesser energy yield
- Electrons lose energy as they pass through the transport system
  - At end of chain, low energy electrons are passed to O_2
  - Simultaneously combine with H^+ to form water

Shows why this is aerobic! No oxygen, no where to pass off electrons

In absence of oxygen, entire mechanism backs up

Chemiosmosis

Discovered in 1961- Peter Mitchell

H^+ gradient builds up as a result of the electrons being passed through the electron transport chain

H^+ are pumped into the intermembrane space, produces a gradient

H^+ flow back to the matrix through protein channels: ATP synthetase

As they flow through the channels, ATP is produced

<u>Energy yield</u>

Net yield from one glucose molecule:

2 ATP from glycolysis

2 ATP from citric acid cycle

32 from NADH and $FADH_2$

Maximum total 36 ATP molecules/1 glucose molecule

In reality, probably less is formed

Other nutrients can be utilized for energy

Fatty acids are broken down into acetyl groups which enter the citric acid cycle

Amino acids are deaminated and enter the citric acid cycle

Energy yield is variable depending on where they enter the steps

<u>Other energy-yielding pathways</u>

Anaerobic respiration

In the absence of oxygen, other molecules serve as the final electron acceptor

Some are obligate (strict) anaerobes, which cannot live in the presence of oxygen

Others are facultative anaerobes, which simply do not use oxygen

Primarily seen in bacteria

Carry out glycolysis as before

End products are CO_2, H_2O and other inorganic compounds

Yield is only 2 ATP molecules/ 1 molecule of glucose

Fermentation pathways

Final electron acceptor is an organic molecule

Carry out glycolysis as before

Yield is only 2 ATP molecules/molecule of glucose

Alcoholic fermentation

$C_6H_{12}O_6$---> CO_2+ ethyl alcohol

Yeasts utilize this pathway, commercially used in baking bread and making beer and wine

Lactate fermentation

$C_6H_{12}O_6$--->lactic acid

Some fungi and yeast carry out this pathway

Utilized commercially to make yeast, sauerkraut
Seen in exercising muscles without sufficient oxygen

Research and Discussion Topics

• Discuss the processes of making beer, wine and bread. What ingredients in these items participate in the fermentation pathways? How are the end products of fermentation utilized?

• Compare and contrast the chemiosmotic process in aerobic respiration, to the similar process seen in photosynthesis.

Teaching Suggestions

• I pass out this study sheet for students to summarize the reactions of aerobic respiration:

	Glycolysis	Conversion of Pyruvic Acid and Citric Acid Cycle	ETS
Where does this occur?			
What are the input molecules?			
What is the carbon output molecule?			
ATP production?			
Electron carriers?			

Suggested Readings

Diamond, J. "The Athlete's Dilemma." *Discover*. August 1991: 78-83. Physiological discussion of the limits on the uses of metabolic energy.

Storey, R.D. "Textbook errors and misconceptions in biology: cell metabolism." *American Biology Teacher*. September 1991. 53 (6): 339-343. Discussion of common areas of misconception and confusion in cellular respiration.

Chapter 9. Capturing Energy: Photosynthesis

Chapter Overview

Nearly all living things depend on photosynthesis. Plants utilize photosynthesis to make sugars to be used in the respiratory pathways, consumers depend ultimately on consuming plants for their food.

Visible light provides the energy for photosynthesis. Wavelengths between violet and red are utilized. Plants tend to appear green because of chlorophyll, which reflects green light, absorbing other wavelengths. Light excites electrons of pigments and they are ultimately expelled and are passed to a series of membrane bound electron acceptors. These redox reactions typically occur within the thylakoid of the chloroplast.

Chloroplasts contain various pigments, primarily chlorophyll *a* and *b*, as well as accessory pigments such as carotenoids. These are housed within the membranes of the thylakoids, which are stacked and called grana. The light dependent reactions occur within the membranes of the thylakoid, the light independent reactions within the ground substance of the chloroplast, the stroma.

The pigments are arranged in clusters called photosystems. Photosystem I contains a reactive molecule known as P700. This specialized chlorophyll molecule is combined with a protein known as the reaction center. Photosystem II contains a similar reactive molecule known as P680. These specialized chlorophyll molecules are surrounded by accessory or antennae pigments pass light energy to the reaction center.

The summary equation for photosynthesis is: Sunlight + 12 H_2O + 6 CO_2 --> 6 O_2 + $C_6H_{12}O_6$ + 6 H_2O The first reactions are the light dependent reactions. Light energy is absorbed by photosystem II and an electron is released to an electron transport system. The original electron is replaced by an electron produced by the splitting of water. The remaining hydrogen ion stays within the thylakoid; oxygen ultimately passes out of the plant. The electron transport system passes the electron to photosystem I, and light energy causes the P700 reaction center to emit another electron, again to a membrane bound electron transport system. This catalyzes the production of NADPH. The hydrogen ion from the splitting of water, and hydrogen ions produced by the electron transport system create a higher hydrogen ion concentration within the thylakoid.

As was seen in aerobic respiration, this hydrogen ion gradient results in the production of ATP. Again, ATP synthetase forms channels in the membrane, through which the hydrogen ions flow, coupling ADP and P_i to form ATP. In contrast to aerobic respiration, this chemiosmosis does not produce ATP for cellular respiration, rather the ATP fuels the light independent reactions.

Carbon dioxide is fixed during the light independent reactions (Calvin cycle). NADPH contributes hydrogen, and ATP from the light dependent reactions permit the step-wise transformation of RP to RuBP to PGA to PGAL, and ultimately to glucose or fructose. This is a partial cycle, as only 1/6 of the PGAL molecules are used to make glucose. The rest are transformed into RP to begin the cycle again. These steps are all catalyzed by enzymes, rubisco being the key. Because the first detectable molecules have 3 carbon atoms, this is referred to as a C_3 pathway. In plants, the glucose and may be joined to form sucrose, a common transport form in plants. Further, starch may be formed as a sugar storage form.

In hot dry weather, plants close their stomates to conserve water. In doing so, carbon dioxide concentrations within the cell decline, and oxygen concentrations rise. In this situation, rubisco binds RuBP to oxygen rather than carbon dioxide, and ultimately the products of this process are carbon dioxide and water. This photorespiration results in a net loss to the plant. Some plants adapted to hot climates are known as C_4 plants, since the first detectable compound formed in carbon fixation has four carbon atoms. This compound is produced in the mesophyll cells, but then transferred to bundle sheath cells, where it ultimately enters the Calvin cycle. These plants experience negligible photorespiration since carbon dioxide is concentrated in the bundle sheath cells, not the mesophyll cells. Other plants utilize a chemically similar process. These CAM plants fix carbon dioxide during the night, when it is cooler, and store that compound (a 4-carbon compound) in the vacuole. Later, it is modified, and it enters the Calvin cycle. Both of these adaptations allow C_4 and CAM plants to inhabit hot dry environments.

Lecture Outline

<u>Photosynthesis</u>

- Critical to life on earth
- Producers fuel the consumers

<u>Solar energy</u>

- Light is a portion of the electromagnetic spectrum
- Gamma waves have very short wavelengths
- Radio waves have very long wavelengths
- Visible light ranges from violet (shorter wavelength) to red (longer wavelength)
- Short wavelengths have more energy per photon; long have more
    - Photosynthesis is based on visible light, with moderate energy levels
    - Visible light excites electrons to higher energy levels
    - When the electron returns to ground state, it may leave the atom
        - Photosynthesis couples reduction and oxidation reactions
- Chlorophyll is the pigment that absorbs light
    - Chlorophyll appears green because it absorbs all visible wavelengths except green
    - Chlorophyll *a* and *b* are the most important in plants

Accessory pigments include carotenoids
Structure of the chloroplast; the site of photosynthesis
Thylakoids: stacks of membranes (grana)
In prokaryotes; thylakoids are extensions of the plasma membrane
Stroma: the fluid within the thylakoids
Some of the reactions happen in the stroma (light independent reactions), some in the thylakoids (light dependent reactions)

Photosynthesis involves formation of energy rich organic molecules from CO_2
Summary:
Sunlight + 12 H_2O + 6 CO_2 --> 6 O_2 + $C_6H_{12}O_6$ + 6 H_2O
Two Stages
Light dependent reactions
Take place only in the presence of light
Chlorophyll absorbs energy and electrons are excited
Excited electrons produce ATP, water is split and $NADP^+$ is reduced
Light independent reactions
ATP and NADPH participate in reactions in which CO_2 is used to make sugars
"Light and dark reactions"
Reactions which occur in the daytime: light dependent and light independent reactions, and aerobic respiration
Reactions which occur in the nighttime: only aerobic respiration

Photosynthesis and photosystems
Photosystems
Aggregations of chlorophyll molecules
Photosystem I contains a reactive chlorophyll molecule known as P700
Photosystem II contains a similar molecule; P680
Photosystems are named for the wavelength of light at which maximum absorption occurs
Accessory pigments: other pigments in the photosystem
Reaction center: protein complexed with P700 or P680
Accessory pigments pass light energy to reactive center
Excited pigment gives an energized electron to an acceptor

Cyclic and noncyclic photophosphorylation
Cyclic photophosphorylation: only PS I is involved
Electrons from PS I return to PS I
Significance is not well understood; not particularly efficient
Noncyclic photophosphorylation: PS II passes electrons to PS I
Electrons in PS II are regenerated from the splitting of water
Called photophosphorylation because light causes the phosphorylation of ATP from ADP

Light dependent reactions

Light energy is absorbed by the pigments

P 680 of PS II emits electrons

Electrons pass to a series of electrons in a series of redox reactions

A proton gradient is produced (similar to aerobic respiration)

Electrons are replaced in PS II by the splitting of water

Water is split into H^+, electrons and oxygen

O_2 is released

ATP is produced

Electrons are ultimately passed to PS I

P 700 of PS I is also excited by light

Electrons from PS I pass to a series of electron acceptors

Ultimately NADPH is produced from $NADP^+$

Chemiosmosis

Proton gradient produced within the thylakoid

ATP synthetase: protein channels

H^+ diffuses through channels, ATP is produced

Function of ATP production: fuel next steps

Light independent reactions

Calvin cycle

Occurs in the stroma

Does not need light

When does it occur? Only in the light

Requirements:

ATP and NADPH from previous reactions

CO_2

Ribulose phosphate (RP), a 5 carbon sugar

Enzymes to catalyze each step

RP is activated by ATP to produced RuBP (ribulose bisphosphate)

Rubisco (an enzyme) combines RuBP and CO_2

RuBP splits into PGA (phosphoglycerate) (a 3 carbon compound)

The Calvin cycle is a C_3 pathway

PGA is converted to PGAL (phosphoglyceraldehyde) with energy from ATP and hydrogen from NADPH

In 6 "turns" of the Carbon cycle, 2 PGA molecules are joined to form glucose or fructose

Remaining PGA molecules are changed to RP to reenter the cycle

Glucose is often subsequently changed into sucrose (transport form of sugar)

Plants may also change it into starch (storage form)

Alternative photosynthetic pathways

Photorespiration

When CO_2 levels are low, rubisco binds RuBP to oxygen, and ultimately the compounds produced are degraded into carbon dioxide and water

This is ultimately a loss to the plant

Photorespiration occurs when plants are water-stressed and close their stomates (CO_2 levels drop, oxygen levels rise)

C_4 pathways

Many tropical plants produce a 4-carbon compound first

This compound is transported to bundle sheath cells

This 4-carbon molecule then proceeds through the Calvin cycle

Photorespiration is negligent

C_4 plants include corn and sugarcane

More efficient in hot, dry climates

CAM plants

Found in crassulacean plants, as well as cactus and pineapple

Plants fix CO_2 during the night, store carbon compound in the vacuole

Later, C_3 pathways process this compound during the day

Research and Discussion Topics

• How quickly does the process of photosynthesis occur? Contrary to what some students think, the light-dependent and light independent reactions both occur only in the presence of light.

• Describe the bacterial processes of photosynthesis. Describe their ecological importance. Where in a lake would you find the photosynthetic purple bacteria? Where would you find the photosynthetic cyanobacteria (blue green algae)?

Teaching Suggestions

• I hand out this review sheet to students to summarize the events of photosynthesis:

	Light Dependent Reactions	Light Independent Reactions
Where occurs?		
Inputs:		
Output carbon molecules:		
Output ATP?		
Output electron carriers?		

• At this point, I also have students compare and contrast the processes of aerobic respiration and photosynthesis:

	Aerobic Respiration	Photosynthesis
What cells utilize this pathway?		
Where do the reactions occur?		
When does this occur?		
Inputs:		
Products:		
Carbon molecules?		
Oxygen?		
ATP?		
Electron carriers?		

Suggested Readings

Chazdon, R.L. and R.W. Pearcy. "Importance of sunflecks for forest understory plants." *Bioscience*. December 1991. 41 (1): 760-766. Plants living on the forest floor utilize short periods of irradiation for photosynthesis.

Eisen, Y. and R. Stavy. "Students' understanding of photosynthesis." *American Biology Teacher*. April 1988. 50 (4): 208-212. A discussion of students' lack of understanding of basic concepts in photosynthesis, and suggestions for improvement.

Storey, R.D. "Textbook errors and misconceptions in biology: Photosynthesis." *American Biology Teacher*. July 1989. 51(5): 271-274.

Hendry, G. "Making, breaking and remaking chlorophyll." *Natural History*. May 1990. 37-40. A description of the global turnover of chlorophyll, relating the fall degradation of chlorophyll seen in temperate zone trees.

Gust, D. and T.A. Moore. "Mimicking Photosynthesis." *Science*. 7 April 1989. 244: 35-40. A discussion of artificial photosynthetic reaction centers.

Friend, D. J. C. "Plant eco-physiology: experiments on crassulacean acid metabolism, using minimal equipment." *American Biology Teacher*. September 1990. 52 (6): 358-361. Various experiments comparing CAM and non-CAM plants, measuring transpiration etc.

Govindjee, and W.J. Coleman. "How plants make oxygen." *Scientific American*. February 1990. 50-58. A description of the "water oxidizing clock."

Chapter 10. Producing a New Generation: Mitosis and Meiosis

Chapter Overview

Organisms have a characteristic chromosome number in their somatic cells, known as the diploid condition (2n). At the end of mitosis, daughter cells also are diploid. Gametes are produced by meiosis, which is a reduction division, and they are haploid (n). Fertilization restores the diploid condition in the zygote. Chromosomes exist in pairs; half of the chromosomes in the zygote are maternal, half are paternal in origin. These members of the pair are known as homologous chromosomes.

The cell cycle is the "life cycle" of the cell. Actual cell division involves separation of the genetic material, known as mitosis, followed by division of the cytoplasm, known as karyokinesis. Prior to cell division, the cell is in interphase. Interphase may be divided into G_1, the primary growth phase, S, the period of DNA synthesis, and G_2, the period of preparation for division.

Mitosis begins with prophase, when the DNA condenses, the nuclear membrane disappears, and complex of microtubules extends from the poles of the cell to the chromosomes. Each duplicated chromosome (sister chromatid) at this point is joined by a centromere. The centromeres are attached via kinetochores to the mitotic spindle. During metaphase, the chromatids are aligned at the equator of the cell. At the end of metaphase, the centromeres divide, and the chromosomes are pulled to the opposite ends of the cell during anaphase. Telophase is characterized by chromosome decondensation, nuclear envelope reformation and the reappearance of the nucleoli. Overlapping with telophase is cytokinesis, division of the cytoplasm. In animal cells, actin and myosin fibers constrict around the equator of the cell, causing the cell to pinch into two. In plant cells, vesicles fuse in the center of the cell forming a cell plate, and ultimately 2 cell walls.

Controls of mitosis include MPF, and hormones like cytokinins; mitosis may be halted by the presence of certain chemicals.

Meiosis is the process which results in the formation of haploid gametes. It may be likened to two rounds of mitosis. Meiosis I begins with prophase I, in which many of the typical events take place, but one of the two unique events of meiosis also occurs. The homologous chromosomes line up and exchange DNA, known as crossing over. This results in many more possible combinations of genetic traits.

In metaphase I, chromosomes are centered on the equator, but the centromeres do not divide, so homologous chromosomes are pulled to the poles during anaphase I. This random movement of the members of the pair is known as independent assortment, and also results in the gametes having varying combinations of chromosomes. Next, in telophase I, chromosomes decondense, and this is typically followed by division of the cell.

In meiosis II, the chromosomes recondense, move to the equator, the centromeres divide and the sister chromatids are pulled to the poles during telophase. At the end, there are four haploid cells produced, each with different combinations of the genetic material.

Lecture Outline

Chromosomes

- Eukaryotes have a fixed number of chromosomes in their somatic cells
- Humans have 46 chromosomes
  - Mitosis is division of somatic cells, which maintains the appropriate number of chromosomes
  - Meiosis is the production of gametes, which is a reduction division
  - Fertilization restores the diploid chromosome number to the zygote
- Chromosomes are paired
  - One set are maternal chromosomes, others are paternal in origin
  - Members of the pair are homologous chromosomes
- The diploid (2n) condition is seen in body cells
- The haploid (n) condition is seen in gametes

Cell cycle of eukaryotes

- Generation time
  - Varies with the species
- Cell division is accomplished by mitosis (division of chromosomes) and cytokinesis (division of the cytoplasm)

Interphase

- The time between divisions
- The G_1 phase (1st gap stage) is characterized by cellular growth
- The S phase is the time of DNA synthesis
- The G_2 phase is characterized by protein synthesis (preparation for mitosis)

Mitosis

- Prophase
  - Chromosomes condense (become visible)
  - Each chromosome contains several cm of DNA condensed into 5-10 μm
  - Chromosomes consist of sister chromatids joined at the centromere
    - Centromeres are attached at the kinetochore to the microtubules
  - Centrosomes duplicate (each contains 2 centrioles)
    - Centrosomes migrate to the poles of the cell
    - Microtubules radiate from the centrosomes (asters)
    - The mitotic spindle of microtubules extends between the poles
    - Chromatids become attached to the microtubules
  - The nuclear envelope disappears

Metaphase
Chromatids are aligned along the equator of the cell
Kinetochores of sister chromatids are attached to spindle fibers which run to opposite ends of the cell
At the end of metaphase, the centromeres divide
During metaphase, chromosomes are very thick and visible under the light microscope; used for karyotyping
Anaphase
Chromatid separation and movement of the chromosomes to the poles
Chromosomes are pulled from the middle, forming a V shape as they are pulled through the cytoplasm
Telophase
Two nuclei reform
Chromosomes decondense
Nuclear membrane reforms, nucleoli reappear, microtubules disappear

<u>Cytokinesis</u>
At the end of telophase, the actual cellular separation occurs
In eukaryotes without cell walls, a cleavage furrow forms
In plants and others with cell walls, a cell plate forms, and ultimately 2 new cell walls
Cell plate materials originate from the Golgi complex

<u>Controls on division</u>
Frequency of mitosis varies with the species, or location within the organism
In humans, skin cells divide constantly, mature nerve cells never divide
MPF (maturation promoting factor) stimulates cells to enter mitosis from G_2
Consists of 2 proteins, cdc2 and cyclin
Cyclin combines with cdc2, forming pre-MPF, ultimately MPF
Other controls include hormones: cytokinins
Colchicine blocks cell division (used to prepare a karyotype)

<u>Meiosis</u>
Reduction division= meiosis
Mitosis v. meiosis
Mitosis
Diploid somatic cells --> diploid somatic cells
1 cell--> 2 cells
Meiosis
Diploid gamete-producing cells --> haploid gametes
1 cell--> 4 cells
There is also genetic exchange during meiosis

Meiosis can be viewed as "2 rounds of mitosis"
Meiosis I and meiosis II

The first meiotic division

Prophase I
- Homologous chromosomes line up= synapsis
- DNA may unwind and pair with other chromosome (the homologue)
  - The 4 are now known as a tetrad
- Importance of crossing over: genetic recombination
  - Introduction of variation in sexually reproducing species
- Total of 4 chromatids for each chromosome pair at this point
- Held together by: centromeres (the sister chromatids) and the points of crossing over (homologues)
- All other typical events of prophase occur as well

Metaphase I
- Chromosomes like up at the equatorial plate
- Centromeres don't divide
- Independent assortment; homologous chromosomes separate (not sister chromatids)
- Very important as source of genetic variability

Anaphase I
- Homologous chromosomes go to each pole

Telophase I
- Nuclei reorganize, cells may divide

Interphase or interkinesis
- No DNA replication, brief phase

The second meiotic division

Mitotic division of the products of meiosis I

Chromatids are not genetically identical
- Because crossing over

Genetic material in the 2 cells is not identical
- Because of independent assortment

Prophase II
- Recondensation of chromosomes

Metaphase II
- Alignment at the equators

Anaphase II
- Movement of chromatids to the poles

Telophase II
- Reformation of nuclei

Cell division= cytokinesis
- End products: haploid cells
- Number: four

Research and Discussion Topics

- Compare and contract mitosis and meiosis. How are they similar? How are they different?

- Discuss the genetic variability introduced by crossing over and independent assortment. How is this important in sexually reproducing populations?

- Define and compare the following, sometimes confusing terms:
  - Chromatid, chromosome, centrome
  - Diploid and haploid
  - Homologous chromosomes, sister chromatics

Teaching Suggestions

- I discuss gametogenesis at this point, mentioning material which will be covered again in the reproduction unit, but it will wake them up. We discuss the site and timing of gametogenesis, and how male and female processes differ.

- I hand out this study sheet comparing the process of mitosis and meiosis:

	Mitosis	Meiosis I	Meiosis II
Events during:			
Interphase			
Prophase			
Metaphase			
Anaphase			
Telophase			
Cytokinesis			
Products:			
Diploid/haploid at beginning?			
Diploid/haploid at end?			

Suggested Readings

Radman, M. amd R. Wagner. "The high fidelity of DNA duplication." *Scienfific American*. August 1988. 40-46. "Quality control" mechanisms in DNA replication.

McIntosh, J.R. and K.L. McDonald. "The mitotic spindle." *Scientific American*. October 1989. 48-56.

Murray, A.W. and M.W. Kirschner. "What controls the cell cycle?" *Scientific American*. March 1991. 56-63. Description of the action of cdc2.

Atkins, T. and J.M. Roderick. "Dropping your genes- a genetics simulation in meiosis, fertilization and reproduction." *American Biology Teacher*. March 1991. 53 (3) 164-169. A lab simulation which models independent assortment, genotypes and phenotypes.

Gonick, L. and M. Wheelis. "Cartoon guide to genetics." 1983. Barnes and Noble Publishers. 124 p. A very well done book describing genetics, including DNA replication and protein synthesis. Applicable to this and following 3 chapters.

Chapter 11. Patterns of Inheritance

Chapter Overview

Prior to Mendel, it was believed that traits were blended, and that traits were directly transmitted from parents to offspring. Through Mendel's laborious experiments, he showed that traits were encoded in "factors" (genes), and that these discrete elements were the basis for inheritance. His work laid the foundation for the work done in the 1900's, which elucidated the part that meiosis plays in gamete formation.

Now it is well known that genes occur in pairs, alternative forms of which are the alleles. Alleles may be the same (homozygous) or different (heterozygous). The gene which is always expressed is the dominant allele, and individuals which have the dominant phenotype may have genotypes of homozygous dominant or heterozygous. The gene which may be masked by the dominant allele is recessive, and the recessive phenotype is a result of a homozygous recessive genotype.

Monohybrid (single allele) crosses may be analyzed using a Punnett square. The parental (P) generation produces the first filial (F_1) generation. When the F_1 generation self-fertilizes, the offspring are the F_2 generation.

When genes are located on different chromosomes, their probability of being inherited is independent. Mendel (fortunately!) worked on characteristics in pea plants which were related to genes on separate chromosomes. The probability of phenotypes and genotypes in a cross involving two traits located on separate chromosomes can be shown in a dihybrid cross. The resulting genotypic ratios are 9:3:3:1. When genes are located on the same chromosome, however, their inheritance is linked, and is not independent. However, due to crossing over during meiosis, their inheritance may not follow simple rules. The further away genes are located (on the same chromosome), the more likely they will be inherited separately, due to crossing over. Looking at the frequency of crossing over gives an idea of gene locations.

Variations on Mendel's simple themes exist. Sometimes traits may be incompletely dominant, such that the heterozygote exhibits traits intermediate to the dominant and recessive traits. In other instances, alleles may be codominant, and the heterozygote exhibits traits of both codominant alleles. The human ABO blood type is an example of a codominant trait, as well as being an example of a multiple allele.

Epistasis refers to the interaction between multiple genes which affect a single trait. Alternatively, one gene may affect more than one trait (pleiotropy). Many morphological features are affected by more than one gene. Height, skin color and other morphological traits are affected by many genes, acting together (polygenes). Most traits which are polygenic show bell-shaped distributions. Finally, expression of some traits are altered by the environment.

Lecture Outline

<u>Mendel's work in pioneering genetics</u>

- Previous ideas included blending inheritance
- Worked on peas- good choice:
  - Large number of true-breeding varieties
  - Perfect flowers
  - Short generation time
  - Easy to grow
- Mendel's experiments
  - Crossed plants with different traits
    - Offspring had dominant trait
  - Offspring self-fertilized
    - Offspring had both dominant and recessive traits, in a 3:1 ratio
- Mendel kept records- important!
  - Presented results- were ignored until the turn of this century
- Interpretation of Mendel's results in modern terminology
  - Traits are transmitted by genes, which occur in alternative forms called alleles
  - Principle of dominance: when dominant and recessive alleles are both present, only the dominant alleles are expressed
  - Principle of segregation: during meiosis, each gamete only gets one copy of each chromosome
  - Principle of independent assortment: during meiosis, chromosomes assort separately and by chance

<u>Modern principles of genetics</u>

- Genes are the unit of DNA which contains the information to code for an amino acid
- In a homologous pair, dominant alleles mask the expression of recessive alleles
- Monohybrid crosses involve the analysis of one allele
  - Punnett squares show all possible combinations
  - The parental generation is denoted P
  - Offspring of the parental generation is called the first filial generation (F_1)
  - Offspring of the next generation is the second filial generation (F_2)
- Genotypes
  - Homozygous dominant, expressing dominant allele
  - Homozygous recessive, expressing recessive allele
  - Heterozygous, expressing dominant allele
- To determine the genotype of an organism with the dominant phenotype, a test cross is performed
  - Cross the individual in question with a homozygous recessive individual

- Dihybrid crosses and gene location
  - Dihybrid crosses involve two different genes which are on different chromosomes
  - Expected genotypic ratios of a test cross are 9 : 3 : 3 : 1
  - Linkage refers to genes located on the same chromosome
    - Typically linked genes are inherited together
    - Crossing over may result in linked genes being inherited separately
    - Studies of linked genes gives information on their location
      - Genes located close together tend to be inherited together
      - Genes located far apart on the same chromosome tend to be affected by crossing over
  - Mapping of the human genome

<u>Variations on Mendel's themes</u>

- Incomplete dominance
  - An intermediate phenotype is seen in heterozygotes
- Codominant alleles
  - Heterozygotes express both alleles
  - Example: human ABO blood types
    - A and B are codominant, O is recessive
    - Possible phenotypes: A, B, AB and O
    - Possible genotypes for type A blood: I^A I^A and I^Ai, o is ii
- Multiple alleles
  - A gene may have more than two alleles in a population (not an individual)
  - Example: ABO blood types
  - Possible alleles: I^A, I^B, and i
- Epistasis
  - Genes may interact, so that one gene masks the expression of another
- Pleiotropy
  - One gene may have more than one effect
  - Example: white (albino) tigers also have defects in vision, and in the heart
- Polygenic traits
  - Many traits are influenced by a multitude of genes
  - Many morphological traits, such as skin color, height, shapes of various body parts (consider the range of nose shapes seen!)
- Environmental effects on phenotype
  - Example: differences seen in "identical" twins, who share identical genetic material

Research and Discussion Topics

• Think about the current state of genetic testing for some very serious conditions, such as Huntington's disease or the inherited form of breast cancer. Would you want to be tested for these genes?

Suggested Readings

Edey, M.A. and D.C. Johanson. *Blueprints, solving the mystery of evolution.* 1989. Penguin Books. Includes an interesting account on Mendel, as well as modern geneticists.

Chapter 12. Human Genetics

Chapter Overview

Early studies of human genetics simply analyzed pedigrees. Today, human geneticists employ molecular techniques to probe into the human genome. We have 23 pairs of chromosomes, containing between 50,000 and 100,000 genes. The autosomes are the first 22 pairs of chromosomes, the last pair are the sex chromosomes. The Y chromosome causes maleness; therefore the genotype XX is female, and XY male. Very few genes reside on the Y chromosome, but the most important gene causes testes development. Many more genes are on the X chromosome, and they are referred to as X-linked genes.

X-linked traits are more common in males than females because males cannot be carriers. Males are referred to as hemizygous, as they possess only one X chromosome. Since most X-linked traits are recessive, the only way a female individual may be produced is by the union of a female carrier (heterozygous) and a male with the particular trait. Examples of X-linked traits include hemophilia, Duchenne muscular dystrophy and red-green colorblindness.

Birth defects may be genetically linked, or environmentally caused. Several are caused by an abnormal number of chromosomes. Down syndrome is caused by nondisjunction, resulting an extra chromosome 21 (trisomy 21). Persons with Down syndrome are mentally retarded and have various physical problems. The genotype of Klinefelter syndrome (male) is XXY; the genotype of Turner syndrome (female) is XO, and each is often sterile, and may show some mental deficits. The XYY karyotype show few differences from the "normal" XY male phenotype.

Other disorders are caused by a single gene, and most are caused by recessive alleles. Phenylketonuria is a condition in which affected persons lack the ability to convert phenylalanine to tyrosine and phenylalanine may build up, causing nervous damage. Newborns are typically tested for this and a special diet during early life may alleviate problems. Sickle-cell anemia is due to one incorrect amino acid in the hemoglobin molecule, and results in a variety of physical problems; capillary blockage, anemia, organ damage etc. One of the most common autosomal recessive disorders in the US is cystic fibrosis. Due to a defect in the chloride channel of the cell membrane, serious respiratory and pancreatic problems develop. In contrast to these disorders, Huntington's disease is an autosomal dominant disease. It is a late acting gene, often showing disease symptoms after reproductive ages, and causes mental and muscular deterioration.

Prenatal screening includes amniocentesis, in which fetal cells from the fluid surrounding the fetus are withdrawn, cultured and karyotyped. Another technique, chorionic villus sampling may be done earlier in pregnancy, but has higher risks. These tests, along with some other enzymatic tests can test for many serious genetic defects. These tests, however pose a variety of ethical questions.

Lecture Outline

<u>Human chromosomes</u>

- 22 pr. autosomes, 1 pr. sex chromosomes
  - XX female; XY male
    - Possession of a Y chromosome determines maleness
    - Y chromosome is very small, contains only information about maleness
- Genes on the X chromosome are X-linked genes
  - X-linked traits are more common in males than females
  - Males are hemizygous (possess only 1 X-linked gene for each trait)
  - Females can be "carriers,", men cannot
  - How can females with X-linked genes be produced?
    - Mother must be homozygous, and father bears the trait
  - Classic example: hemophilia

<u>Human genetic disorders</u>

- Birth defects may be inherited, others environmentally influenced
- Down syndrome is caused by a trisomy of chromosome 21
  - One of few viable trisomies
  - Show mental retardation, susceptibility to diseases and cancer
  - More common with increasing maternal age
- Klinefelter syndrome: XXY
  - Phenotypically male, may be mentally retarded
- Turner syndrome: XO
  - Lack a second sex chromosome
  - Phenotypically female
  - Sterile, may be retarded
- XXY karyotype: phenotypically male, few other effects
- Phenylketonuria (PKU)
  - Lack enzyme that converts phenylalanine to tyrosine
  - Must be put on low phenylalanine diet
  - Avoid Nutrasweet
- Sickle-cell anemia
  - One incorrect amino acid in hemoglobin
  - Leads to a variety of physical problems
    - Sickled red blood cells block capillaries, anemia
  - Most common in persons of African descent
- Cystic fibrosis
  - Most common serious genetic disorder among Caucasians
  - Malfunctioning chloride ion channel in cell membranes
  - Results in respiratory, pancreatic problems
- Huntington's disease
  - A dominant autosomal disorder
  - Late acting gene
  - Leads to mental, muscular deterioration
  - Gene was identified in 1993

Prenatal testing
- Amniocentesis
  - Removal of fluid (amniotic fluid) surrounding the fetus containing fetal cells
  - Cells are cultured, karyotyped
- Chorionic villus sampling (CVS)
  - Removal of sample from the chorion
  - May be done earlier, but has higher risks

Research and Discussion Topics

- What genetic disorders are currently identifiable through prenatal testing (chorionic villus sampling or amniocentesis? How do these procedures work?

Suggested Readings

Offner, S. "A plain English map of the human chromosomes." *American Biology Teacher* . February 1992. 54 (2): 87-91. A written description and an illustrative map of many humans genes which have been identified.

Offner, S. "A revised map of the human chromosomes." *American Biology Teacher*. October 1993. 55 (7): 406-410. An update.

Barnes, D. M. "Fragile X syndrome and its puzzling genetics." *Science*. 13 January 1989 243: 171-172. An example of an X-linked gene which is not always expressed in males, but is a cause of many cases of mental retardation (2nd only to Down syndrome).

Glausiusz, J. "The year in genes." *Discover*. January 1994. 92. A one-page description of 9 genes that were identified in 1993.

Palca, J. "The promise of a cure." *Discover*. June 1994. 76-86. Current gene therapy experiments on cystic fibrosis.

Diamond, J. "The return of cholera." *Discover*. February 1992. 60-66. A description of the physiology of cholera, including the link to cystic fibrosis.

Revkin, A. "Hunting down Huntington's." *Discover*. 99-108. A description of Nancy Wexler's work on Huntington's disease.

Chapter 13. DNA: The Molecular Basis of Inheritance

Chapter Overview

The first experiments that bridged the gap between Mendelian genetics and modern molecular genetics had several questions to answer? What is the genetic material? Where is it stored in the cell? How is it reproduced with such accuracy? The first experiments to yield answers to some of these questions was carried out by Hammerling on *Acetabularia*, a macroscopic single celled alga. It was found that the portion of the cell which contained the information necessary for regeneration was in the holdfast, which was also the site of the nucleus.

Experiments by Griffith indicated that there was some interaction between nonvirulent, live bacteria, and dead virulent bacteria that allowed the combination of the two to kill mice. His experiments were furthered by Avery, and it was ultimately found that DNA was the "transforming principle." However, it was not for several years that experiments by Hershey and Chase supported the idea that DNA was the genetic material. Hershey and Chase experimented with bacteriophages, and radioactively labeled viruses showed that DNA was the genetic material (not protein, as was earlier believed).

X-ray diffraction patterns done by Wilkins and Franklin in the early 1950's gave clues that the DNA molecule was helical, and long and thin, with repetitive units. Francis Crick and James Watson finally deduced the correct structure of DNA in 1953, described in a single-page paper in the British journal Nature.

DNA is now known to be a double helix, like a twisted ladder, with phosphates and pentose sugars (deoxyribose) forming the sides of the ladder, and nitrogen-containing bases forming the rungs. The bases in DNA are adenine, thymine, guanine and cytosine. Adenine and guanine are single-ringed structures (purines), and thymine and guanine are double-ringed structures (pyrimidines). Purines bond to pyrimidines, making the diameter of the DNA molecule uniform. This bonding principle is known as complementarity. Based on the base sequence of one side of the molecule, we can predict the base sequence on the other side. The two strands are antiparallel, meaning that the molecular structure of the phosphate and sugar molecules are opposite in the two sides.

Watson and Crick noticed that the structure of the DNA molecule suggested a possible copying mechanism. We refer to the mechanism of DNA replication as semiconservative. Enzymes separate the bases of the molecule, and add new bases in a complementary fashion. DNA polymerases are responsible for adding bases at appropriate spots. Therefore, each daughter molecule is half parental DNA, half new DNA. Enzymes "check" the growing DNA molecule for errors, and will excise and replace any inappropriately paired bases. These mutations are random and are not necessarily harmful, as they are the raw material for evolution.

DNA is organized in chromatin, and is highly organized when in the form of chromosomes. DNA is wrapped around proteins called histones. Histones associated with DNA are referred to as nucleosomes. Further looping associated with scaffolding proteins produces chromatin. Chromosomes are even more highly organized.

Lecture Outline

The early genetic experiments

- Hammerling and *Acetabularia*
  - Only the portion containing the nucleus regenerated (holdfasts)
  - If the stalks of two different species *Acetabularia* are switched, the regenerated cap is similar to the species of the holdfast
- Griffith and Avery
  - S strain of *Streptococcus pneumoniae* was pathogenic
  - R strain is not pathogenic
  - Experiment:
    - S strain injected into mice kills them
    - R strain injected into mice does not kill them
    - Heat killed S strain injected into mice does not kill them
    - Heat killed S strain + live R strain injected into mice kills them!
  - How? Griffith called it the transforming principle
  - Griffith was killed by a bomb in England accidentally during WW I
  - Avery (Rockefeller Inst. in US) carried out further experiments
    - Found that purified DNA would give them same results
    - Results were not widely accepted
- Hershey and Chase
  - Worked on bacteriophages (phages); viruses that attack bacteria
  - Radioactively labeled viral proteins with ^{35}S, labeled DNA with ^{32}P
  - Added viruses to bacteria, bacteria became radioactive with ^{32}P, not ^{35}S
- Wilkins and Franklin
  - X-ray diffraction gave clues as to the structure of DNA
  - Believed that DNA was long and thin and helical
- Crick and Watson
  - Described the correct molecular structure of DNA in a one-page paper in Nature in 1953
  - Nobel prize in 1962, along with Wilkins (Franklin had died of cancer in 1958)

Structure of DNA

- Backbone of alternating phosphate and pentose sugars
- "Rungs" of the ladder are nitrogenous bases: adenine, thymine, cytosine and guanine
  - Adenine bonded to thymine, cytosine bonded to guanine
- Double helix

DNA is a polynucleotide of nucleotide monomers
Nucleotide structure:
- Pentose sugar
- Phosphate group
- Nitrogen containing base
  - Cytosine and thymine are pyrimidines; single ring structures
  - Adenine and guanine are purines; double ring structures
- The order of the bases is the genetic code

DNA is double stranded; the two strands are complementary
- A bonds to T; C bonds to G
- A double ringed base always bonds to a single ringed base
- Diameter of molecule is uniform

The two strands run in opposite directions (due to arrangements of their pentose sugars); they are antiparallel

DNA replication

Because the strands are complementary, each strand may be used as a template for the other

Replication occurs during the S phase of interphase
- The DNA molecule "unzips," and each side is used as a template to make a new molecule

Replication is semiconservative; each daughter strand is half new, half old

Where the bases of the DNA molecule are separated is a replication fork
- Multiple replication forks speeds replication

Enzymes involved include DNA polymerase which adds new bases
- DNA polymerase is responsible for bringing the correct base to be paired

Mutations occur when errors occur in DNA replication
- Mutations are not necessarily bad- they are the raw material for evolution
- Cells contain enzymes which "check" for errors and correct them

DNA organization

DNA is organized around proteins called histones

Histones with DNA wrapped around them are nucleosomes

Chromatin is a DNA/protein complex
- Loops are held together by other proteins called scaffolding proteins

Research and Discussion Topics

- Describe the levels of organization in the DNA molecule when in the form of a chromosome. How does this differ from the chromatin form? Why do the differences exist?

Teaching Suggestions

• I love the inclusion of quotes from "What Mad Pursuit", by Crick. There's such a magic to the quest for the structure of DNA. I also read the beginning and the end paragraphs from their paper in Nature. Such classic lines, such as "The specific pairing we have postulated immediately suggests a possible copying mechanism for the genetic material." What an understatement!

Suggested Readings

Crick, F. "What mad pursuit." 1988 Basic Books Publishers. 182 p. A personal view of his life.

Watson, J.D. "The double helix." 1980. W.W. Norton Publishers, New York. 298 pages. A collection of papers relating to the discovery of the structure of DNA.

Roberts, L. "Chromosomes: the ends in view." *Science*. May 1988. 240: 982-983. A description of telomeres.

Chapter 14. Gene Function: RNA and Protein Synthesis

Chapter Overview

Genes carry the information to assemble polypeptide chains in the triplets of bases in the DNA molecule. DNA is used as a template from which mRNA is made, which carries the information of the sequence of the bases in its codons. These codons code for amino acids; the code is redundant, and some codons are read as "stop" and "start" codons. This code is nearly universal, indicating the shared evolutionary heritage of all organisms.

During the process of transcription, mRNA is assembled along one strand of the mRNA. It copies the length of the gene with the aid of RNA polymerases, after which introns are cut out, leaving a completed mRNA transcript of spliced exons. The mRNA strand then travels to the cytoplasm, and joins with the two subunits of the ribosome.

The ribosome is composed of proteins and rRNA, and is the site of polypeptide chain assembly. During the process of translation, the mRNA molecule moves along the ribosome, pairing codons of the mRNA with complementary tRNA anticodons. Each tRNA molecule carries an appropriate amino acid to the ribosome, and enzymes catalyze the joining of the amino acids into a polypeptide chain. Multiple copies of proteins may be copied from one mRNA molecule to speed the amount of protein synthesized.

Mutations are changes in the DNA; they occur spontaneously due to errors in replication, due to jumping genes or due to mutagens, such as X rays and UV radiation.

Lecture Outline

<u>Purpose of genetic material</u>
- Protein synthesis
- Protein functions:
  - Respiratory processes
  - Active transport
  - Cell movement
  - Enzymes
- Genes carry the information to build a polypeptide
- Proteins are not built on the DNA molecules
  - Protein synthesis involves RNA, DNA, ribosomes and enzymes
  - Amino acids are the building blocks of proteins
  - Amino acids are built on the RNA transcript
- Transcription
  - DNA unwinds; RNA copies one of the strands
  - The "language" of transcription is the bases on both DNA and RNA

Translation

Occurs on the ribosome

Amino acids are assembled into polypeptide chains

Translation changes the "language" of RNA (bases) to the "language" of proteins (amino acids)

The genetic code

DNA has four bases; must be read in triplets

Triplet code has sufficient combinations for 64 amino acids; therefore code is redundant (equivalent)

Read in groups of 3 consecutive nucleotides

Each block of 3 nucleotides codes for 1 amino acid

Several triplets mean stop copying

Code nearly universal

Universality is what allows genetic recombination

RNA and transcription

RNA has ribose as its sugar

Uracil replaces thymine; bonds to adenine

Three kinds of RNA

Messenger RNA (mRNA) copies and carries genetic instructions from DNA

Ribosomal RNA (rRNA) and proteins compose the ribosome

Transfer RNA (tRNA) brings amino acids to the ribosome

RNA is made by the process of transcription

Similar to the process of DNA replication

DNA molecule unwinds; only one strand is copied

Enzyme is RNA polymerase

Transcription begins at a promoter

Proceeds in a 5′ to 3′ direction

Bases pair; C to G; A to U

Transcription stops when it comes to a stop signal

mRNA molecule falls from the DNA strand

Each sequence of three bases copied from the DNA triplets is known as a codon

Introns are cut out of the mRNA molecule; exons are spliced together

Translation; protein synthesis

Translation= transfer RNA (tRNA) and ribosomal RNA (rRNA) and mRNA together construct proteins

Ribosomal RNA and proteins form the ribosome

Has two subunits; one large and one small

tRNA is a small molecule of RNA

The anticodon is a recognition site of 3 exposed bases

This site will ultimately bind with the mRNA codons

Amino acid attachment site

Stages of translation

Initiation stage; mRNA attaches to the small subunit of ribosome

Large subunit then is joined to the small subunit

AUG sequence on mRNA signals the beginning of the message

Polypeptide chain (protein chain) elongation

tRNA anticodons pairs with complementary mRNA codons

tRNA contributes its amino acids to the growing polypeptide chain

P site is the position where amino acids are already attached

A site is the position of the next mRNA codon to be presented to the tRNA

The mRNA molecule shifts so that the codon in the A site shifts to the P site

The tRNA molecule in the P site is released

Termination

Some codons do not code for any amino acids

Trigger release factors to detach the protein chain and the mRNA from the ribosome

mRNA may be used to make multiple copies of a protein

Multiple ribosomes attach; called polysomes

<u>Mutations</u>

Changes in the DNA result in changes in the protein product

Some occur spontaneously from errors in replication

Transposons ("jumping genes") move spontaneously within a chromosome or between chromosomes

Often causes mutations

When transposons move out of a gene, can reverse the mutation

Discovered by Barbara McClintock; Nobel Prize in 1983

Mutagens include ionizing radiation and various chemicals

X rays or gamma rays

UV in solar radiation

Linear-dose relationship

Smaller dose; smaller mutagenic effect

Even smallest dose causes some mutations

Research and Discussion Topics

• Research McClintock's work on transposons, which led to her Nobel Prize. How was corn involved in her research?

• Discuss the mutagenic effects of X-rays and UV radiation. What do they do, at the molecular level?

Teaching Suggestions

• To illustrate the difference between transcription and translation, I use this example. Before the advent of printing presses, monks spent long hours transcribing books, primarily bibles. They transcribed from the English language to the English language (same language). However, if you were to translate a book, you change the language; for example from English to Spanish.

• I have found that nothing beats an animation sequence to illustrate protein synthesis. No matter how effective your drawings or transparencies are, a visual presentation is a must. Try to purchase some visual aid!

• Add "problems" to your lecture, particularly near the end to cement student's understanding. Start with simple problems like "What mRNA would be copied from the DNA sequence ATACGG?" Move through transcription and translation, and end with more difficult problems like "What DNA sequence would code mRNA to which the tRNA anticodon is AUU?" (Answer: ATT).

• When discussing the base language of DNA and how mutations affect it, I use examples from the English language. For example, consider the sentence: "

"ANDTHECATANDDOGEATANDRUN"

It can be read in a sequence of consecutive, unpunctuated three-letter words. I write it on the board, and read it aloud. Then I show them that if a mutation caused the letter C to be replaced by a R, it does cause a change in meaning, but not a drastic one (base substitution mutation). However, a base deletion or addition, or a jumping gene can cause a drastic alteration of the message. I show them that if the "C" is removed, the message becomes AND THE ATA NDD OGE ATA NDR UN, making no sense. Read it aloud; students will see that it is garbled!

• The following is a worksheet that I give my students to allow them to understand this process more clearly. I offer it as extra credit to be turned in on the day of the exam. On this exam this year, I found that students who completed this extra credit project scored significantly higher on the exam. This was modified from: Offner, S. "Teaching Biology around themes: Teach proteins and DNA together." *American Biology Teacher*. February 1992. 54(2): 93-101.

This is the base sequence on one strand of a certain DNA molecule:

A A T G C C A G T G G T T C G C A C

1. Give the base sequence on the complementary DNA strand (the other side).

2. Give the base sequence of the strand of mRNA which would be constructed from the original DNA strand.

3. What amino acid sequence would this mRNA code for?

4. If the fourth nucleotide of the original strand of DNA were changed from G to C, what would the resulting mRNA strand read?

5. What would the resulting amino acid sequence be?

6. If a G were added to the original DNA strand after the 3d nucleotide, what would the resulting mRNA read?

7. What would the resulting amino acid sequence be?

8. If the 8th nucleotide on the original DNA strand were changed from G to C, what would the resulting mRNA strand read?

9. What would the resulting amino acid sequence be?

10. Consider the following DNA base sequence:

C A C G T G G A C T G A G G A C T C C T C

What is the mRNA sequence that would result from this DNA sequence?

11. What is the amino acid sequence that would result from this mRNA sequence?

12. If the 17th nucleotide in the original DNA molecule were changed from T to A, what mRNA would the new DNA code for?

13. What is the amino acid sequence that would result from this mRNA sequence?

This is the change that results in sickle cell anemia. Using your text book as a reference, describe the occurrence of sickle cell anemia, the phenotype and genotype of the disease and the relationship of sickle cell anemia and malaria.

Suggested Readings

Flannery, M.C. "Respect for RNA." *American Biology Teacher*. October 1991. 53 (7): 438-441. RNA as enzyme, RNA editing, the "RNA world."

Radetsky, P. "Genetic heretic." *Discover*. November 1990. 79-84. A description of Thomas Cech's work on RNA, culminating in the Nobel Prize in 1989.

Chapter 15. Gene Regulation

Chapter Overview

In an adult human, trillions of cells do a variety of functions. Since all cells originated from one fertilized egg cell, nearly all cells share the same genetic material. However, cells of the follicles in your scalp produce keratin (hair), and certain cells of the pancreas produce hormones like insulin; therefore these cells must be controlled so that they only transcribe the appropriate genes coding for keratin, or insulin. Most genes are not constitutive, and must be turned on and off as needed.

Studies of *E. coli,* pioneered by Jacob and Monod showed that bacteria normally produce enzymes to metabolize glucose, but the genes to digest lactose are organized in a functional unit, the *lac* operon. The lac operon is an inducible operon, and consists of structural genes, a promoter, an operator and a regulator gene. In this operon, the regulator gene produces a repressor, which binds to the operator, which overlaps the promoter, and thereby prevents transcription. However, in the presence of lactose, the repressor cannot bind to the operator and transcription proceeds. Inducible operons typically code for enzymes involved in metabolism of nutrient molecules, and are turned on only when needed.

Other operons are repressible, and are normally turned on. The *trp* operon also consists of the same basic parts, but the end product, tryptophan acts to inhibit transcription by binding to the repressor, causing it to bind to the operator, which also overlaps the promoter. Repressible operons typically code for the synthesis of essential molecules such as amino acids, nucleotides, etc.

Eukaryotes have different modes of genetic control, and these include transcriptional controls, mRNA processing, translational controls, post-translational controls and feedback inhibition.

Lecture Outline

<u>Gene expression</u>

- Nearly all cells of an organism contain identical genetic material
- Cells only need to produce certain proteins at a given time
- Gene regulation best understood in prokaryotes
    - Many studies on *E coli*
- Constitutive genes
    - Code for proteins that are continually needed
        - Example: enzymes needed for glycolysis
- Most genes are switched "off"
- Very "hot" topic in science research currently

Control mechanisms in prokaryotic genes

- Transcriptional controls
  - Allows bacteria to make whatever enzymes are needed at a particular time
  - Research by Jacob and Monod
    - *E. coli* exposed to glucose (monosaccharide) and lactose (disaccharide)
    - First consume glucose, lactose later
      - Time delay due to need to produce enzymes to digest lactose
- The operon
  - Clusters of genes
  - Four parts:
    - Structural genes which code for the enzymes
    - Promoter which is where RNA polymerase first binds
      - Repressor molecules may bind to promoter and block it
    - Operator which acts as a control switch (on or off)
    - Regulator gene which controls the operator
  - *Lac* operon
    - Three structural genes coding for enzymes which transport lactose into the cell and ultimately digest lactose
    - Regulator gene is constitutive; always turned on
    - Repressor normally binds to operator, blocking the promoter
    - Under normal conditions, the gene is not transcribed
    - When lactose is present, the repressor cannot bind to operator
      - Promoter is unblocked, RNA polymerase binds and transcribes structural genes
    - Inducible system: normally turned off, but presence of substrate turns it on
    - Most inducible operons are involved in breakdown of nutrient molecules
- *Trp* operon
  - 5 structural genes which produce tryptophan
  - Also contains a regulator gene, operator and promoter
  - Operator overlaps the promoter
  - Regulator gene codes for a repressor protein that binds to the operator and therefore blocks the promoter
  - When tryptophan is present, it binds to the repressor and thegether bind to the operator
  - The *trp* operon is a repressible operon; the end product blocks transcription
    - Normally, the *trp* operon is turned on

Genetic control in multicellular eukaryotes

- During cell division in the embryo, cells differentiate
- Eukaryotic genes are not typically organized in operons

5 types of control of gene expression:
Transcriptional controls
 Chromatin has varying levels of condensation
  Condensed DNA cannot be transcribed
 Enhancers increase the rate of transcription
mRNA processing
 Introns are cut out; exons are spliced together
Translational control
 Varying numbers of protein molecules may be copied from one mRNA molecule
 Some mRNA molecules are unstable and degrade rapidly
Post-translational processing
 Enzymes may be transformed to an inactive form
Feedback inhibition
 Seen in complex metabolic pathways
 End products may inhibit earlier reactions

Research and Discussion Topics

• Most of your students will have either seen the movie Jurassic Park, or read the book. Discuss the probability that little bits of DNA taken from Mesozoic insects could yield an entire dinosaur. Relate the problems in such a task to the genetic controls discussed in this chapter.

• Have students research the principle of totipotency. At what point does the human embryo lose totipotency?

• Research the varying levels of organization of DNA in the form of chromatin and chromosomes. Relate these varying levels of "packing" to transcriptional controls.

• Describe the function of Barr bodies. How are the X chromosomes (Barr bodies) genetically controlled?

• Much current research has to do with homeoboxes, which are controlling genes in the development of eukaryotes. Describe the current state of knowledge concerning homeoboxes.

Teaching Suggestions

• Operons tend to confuse students. I find it useful to draw them on the board, color coding the various parts. Put up both the *lac* and the *trp* operons and compare and contrast their structure and function.

Suggested Readings

Caldwell, M. "How does a single cell become a whole body?" *Discover*. November 1992. A description of the homeobox and gene regulation.

Pines, M. Editor. "From egg to adult." A report from the Howard Hughes Medical Institute. Available (free!) from the Howard Hugues Medical Institute, 6701 Rockledge Dr, Bethesda, MD 20817. 55 pages. A complete, compelling description of gene regulation and developmental biology.

Rennie, J. "DNA's new twists." *Scientific American*. March 1993. 122-132. A discussion of some recent discoveries in the area of genetics, including jumping genes, gene imprinting, and directed mutations.

Weintraub, H.M. "Antisense RNA and DNA." *Scientific American*. January 1990. 40-63. Genetic controls of RNA molecules.

Steitz, J.A. "Snurps." *Scientific American*. June 1988. 56-63. A description of small nuclear ribonucleoproteins, which remove introns from mRNA molecules.

McKnight, S.L. "Molecular zippers in gene regulation." *Scientific American*. April 1991. 54-64. Leucine links (zippers) between polypeptide chains and genetic controls.

Gillis, A.M. "Turning off the X chromosome." *Bioscience*. March 1994. 44 (3): 128-132. Mechanisms of X chromosome inactivation.

Chapter 16. Out with the Bad Genes, in with the Good

Chapter Overview

Selective breeding is the basis of our agriculture, and is getting a boost currently by biotechnology. Biotechnology is based on genetic engineering, in which DNA from a parental cell is altered, typically combined with a gene (DNA) of interest, and then inserted into the same or a different cell. An important tool in recombination are plasmids, circular strands of DNA found in bacteria. This process is aided by enzymes known as restriction enzymes which cut in specific ways, at palindromic spots in the DNA sequence. Complementary ends of the DNA of interest, and the cut ends of the plasmid pair with the aid of DNA ligase. Plasmids are the most common vectors, but viruses and gene guns may also be used. Viruses and gene guns may be used to insert recombinant DNA into mammalian cells.

Using plasmids as vectors, the host bacteria are treated with heat and calcium to promote plasmid uptake. They are then referred to as transformed. To identify the transformed cells which received the entire correct gene, bacteria are cultured, then treated with antibiotics to kill cells which were not transformed. Ultimately, genetic probes are used to identify the cells of interest. Further, to ensure expression, genes which are inserted must include the regulatory and promoter genes; eukaryotic genes are inserted in the form of cDNA, so noncoding introns are not introduced.

Polymerase chain reaction is a technique of alternate treatments of heat and enzymes which allow production of multiple copies of DNA.

Current research in biotechnology holds the promise of developments to improve human health and promote agriculture. Currently, a variety of products are produced by bacteria, such as human insulin, human growth hormone, factor VIII and tPA. Future genetic therapy may allow insertions of normal genes in defective cells to correct human genetic disorders. Also, genetic manipulation of both plants and animals allows them to be resistant to diseases or have higher production. The crown gall bacterium, *Agrobacterium*, is commonly used as a vector in research on plants, but only is effective on dicots, and ones that can be grown by tissue culture. Various regulations on genetic manipulations ensure that potentially hazardous strains are not introduced into the environment.

Lecture Outline

<u>Biotechnology</u>

- Selective breeding has been used for thousands of years
- Biotechnology's goal is to improve human health and agriculture
- Genetic engineering is the modification of DNA of a living organism
    - Same goals as selective breeding
    - Much faster process

Recombinant DNA technology

- Began in the 1970's
- Foreign DNA is inserted into bacteria or eukaryotes
  - Genes are then expressed
  - Bacteria transfer genetic material naturally
- DNA must be isolated from "parental" cell
  - Utilize restriction enzymes
    - Cut in short palindromic spots
    - Restriction enzymes normally protect bacteria from viruses
  - Restriction enzymes leave "sticky ends", which can spontaneously pair with other complementary ends cut by enzyme
  - Splicing is accomplished with DNA ligase
- Vectors carry the DNA
  - Three types of vectors: plasmids, viruses and gene guns
  - Plasmids are small circular DNA molecules in bacteria
    - Are accessory DNA rings
    - Plasmids are cut with restriction enzymes, gene of interest is inserted.
    - Utilized in inserting genes into bacteria and plant cells
  - Viruses introduce recombinant genes into mammalian cells
    - Viruses are inactivated so they do not cause disease
  - Gene guns shoot DNA coated pellets into live cells
    - Work on single cells, plant and animal cells
- Recombinant techniques using plasmids
  - Plasmids are cut using restriction enzymes
  - Plasmids are the vectors for desired genes
  - Bacteria are treated with heat and calcium so it will take up recombinant plasmid
    - Bacteria are then referred to as transformed
  - Plasmid must be amplified
    - Since plasmids carry information which allows them to be resistant to antibiotics, they are treated with antibiotics
    - Only bacteria which took up the plasmid survive
    - Surviving bacteria are cultured and cloned
    - Genetic probes allow detection of bacteria which have the gene of interest
      - Probes are radioactively labeled segment of RNA or DNA
      - Radioactive bacterial colonies are detected
    - Bacteria which contain the gene of interest must express that gene
      - In regulated genes, associated regulatory and promoter genes must be included
      - Inserting cDNA (complimentary DNA) allows the desired protein to be synthesized

PCR

- Polymerase chain reaction allow production of large amounts of DNA to be produced
- Treatment of heat separate the DNA molecule
  - Enzymes then replicate the two strands
  - Further treatments of heat and enzymes double the amount of DNA with each treatment

Biotechnology; present and future

- Production of insulin was one of the first commercial products
- Human growth hormone to treat pituitary dwarfism
- Factor VIII to treat hemophiliacs
- Treating human genetic defects
  - Approximately 3500 genetic disorders in humans
  - Gene replacement therapy
    - Current research on ADA, SCIDS etc.
- Genetic research and agriculture
  - Genetically engineered cows are resistant to rinderpest
  - Improvements in the animals to increase production
- Biotechnology and plants
  - Less restrictions on experimentation on plants
  - Corn resistance to the European corn borer
  - Much research using *Agrobacterium tumefaciens*
    - Crown gall bacterium; used as vector in plant research
    - Limitations
      - Only works on dicots
        - Other techniques must be used on monocots like corn, or wheat
      - Must use tissue culture techniques
        - Not developed for all plants
      - Only can manipulate DNA in nucleus, not chloroplasts

Safety and biotechnology

- Stringent guidelines
- No problems so far
- Benefits far outweigh possible hazards

Research and Discussion Topics

• Discuss current advances in testing for genetic conditions which have been made possible using genetic technology. For example, a test which is reasonably reliable allows people to find out whether they have the gene for Huntington's disease. Would you be tested?

• Research the various regulations in the US on genetic experimentation. Compare and contrast the regulations using prokaryotic, plant and animal genetic manipulation.

• Recently, a genetically engineered tomato has appeared on the market. Find out how the tomato was created.

• Research various careers in biotechnology. What educational requirements are necessary for these jobs? Where in the US are most of these biotechnological companies located?

• What is the current status of the patent law associated with biotechnology? Can you get a patent for a genetically engineered bacterium? A plant? An animal, like a genetically altered mouse?

• What is the current legal status of allowing DNA evidence in the courtroom? Investigate current trials which have involved DNA evidence.

Teaching Suggestions

• Discuss current legal and illegal uses of genetically engineered human growth hormone as used by pituitary dwarfs, and body builders, respectively. What are the ethical considerations concerning the use of HGH? How might the inappropriate use of HGH lead to acromegaly?

Suggested Readings

Hepfer, C.E., J.B. Piperberg, G.M. Farganis. "An introduction to DNA fingerprinting." *American Biology Teacher*. April 1993. 55 (4) 216-221. A description of the methodology and applications of DNA fingerprinting.

Hunkapiller, T., R.J. Kaiser, B.F. Koop, L. Hood. "Large-scale and automated DNA sequence determination." *Science*. 4 October 1991. 59-67. Description of DNA sequence analysis, and future technologies.

Gasser, C.S. and R.T. Fraley. 'Transgenic crops." *Scientific American*. June 1992. 62-69. An excellent description of biotechnological advances in agriculture.

Grady, D. "The ticking of a time bomb in the genes." *Discover*. June 1987. 26-37. A description of Huntington's disease and Nancy Wexler's research.

Baskin, Y. "DNA unlimited." *Discover*. July 1990. 77-79. A short, easy-to-read description of PCR technology.

Neufeld, P.J. and N. Colman. "When science takes the witness stand." *Scientific American*. May 1990. 46-53. A description of forensic DNA methods.

Lowenstein, J.M. "Whose genome is it anyway?" *Discover*. May 1992. 28-31. The human genome project's aim is to map a composite of hundreds of individuals.

Goodman, B. "The genetic anatomy of us (and a few friends)." *Bioscience*. July/August 1990. 40 (7) 484-489. Research on *C. elegans, E. coli* and *Drosophila* aids in the human genome project.

Gillis, A.M. "Therapeutics thrust of transgenics." *Bioscience*. December 1992. 42 (11): 115-116. Transgenic animals produce products like hemoglobin for humans.

Beardsley, T. "Big-time biology." *Scientific American*. November 1994. 90-97. A description of the boom in biotechnology as a science and a business.

Murray, T. "The growing danger from gene-spliced hormones." *Discover*. February 1987. 88-92. Possible inappropriate use of genetically produced hormones, like human growth hormone.

Erlich, H.A., D. Gelfand and J.J. Sninsky. "Recent advances in the polymerase chain reaction." *Science*. 21 June 1991. 1643-1350.

Chapter 17. Darwin and Natural Selection

Chapter Overview

Naturalists over the ages have noted the variety of living things, yet the obvious similarities between evolutionarily related organisms. Most people, however, believed that species were created by a creator, looking similar to present-day forms. Lamarck proposed the first theory of evolution, but his <u>mechanism</u> was incorrect. The mechanism of natural selection was proposed jointly by Wallace and Darwin in 1858. Darwin formulated his ideas as a result of years of research and travel, Wallace came up with similar ideas in a shorter period of time. During the several decades following the publication of the "Origin of Species," there was much consternation because Darwin's theory was at odds with the biblical account. During the 1920's, the synthetic theory of evolution merged natural selection with population genetics. We now know that mutations provide the raw material for variation, and variation results in natural selection, differential reproduction and changes in the allelic proportions in a population.

Evidence supporting the evolutionary theory comes from many sources. Fossils show the changes in organisms over time, although the fossil record is relatively incomplete, and is biased toward aquatic organisms with hard parts. Comparative anatomy shows similarities between organisms which have common ancestors (homology) and convergent evolution of structures seen in organisms which do not share ancestors (analogy). Vestigial structures also give clues. The study of the distribution of living things, biogeography, gives clues to the movement of the continents and the center of evolutionary origin of taxonomic groups. Closely related groups also have similar developmental patterns, similar amino acid sequences, similar DNA sequences, and nearly all origins share the same genetic code, indicating that all organisms evolved from a common ancestor.

Lecture Outline

<u>Evolution</u>

- A change in living things
- Darwin: "Descent with modification"
- Change in frequency of alleles in a population

<u>Historical view of evolution</u>

- Aristotle recognized diversity, thought species moved to a more perfect state
- Fossils influenced early naturalists, including da Vinci
- Lamarck, 1809
  - First coherent theory of evolution (however, it was wrong)
  - Inheritance of acquired characteristics
  - Organisms evolved toward greater complexity

Darwin
- 5 year voyage on H.M.S. Beagle
- Charted coastline of S. America, collected many specimens, made many observations
- Galapagos Islands

Malthus; economic theory
- Populations increase geometrically, food supply increases arithmetically
- Population growth should theoretically outstrip food supply
- Ideas influenced Darwin

Lyell
- Geologist, also influenced Darwin

Darwin spent 20 years studying a variety of things, writing books

Alfred Wallace was the catalyst to publish
- Studied in SE Asia and came up with similar ideas; sent them to Darwin

Darwin and Wallace presented papers to the Linnaean Society, July 1858

"On the Origin of Species..." published in 1859
- Natural selection as the mechanism of evolution

Four major ideas
- Overproduction
  - Ideas from Malthus
  - Populations increase geometrically
- Variation
  - Must be heritable variation
  - Darwin did not know mechanism of genetics
- Limits on population growth
  - Due to competition for scarce resources
- Survival to reproductive age
  - Colloquially called "survival of the fittest"
  - Better to point out that it is "reproduction of the fittest"
- Natural selection results in changes in alleles within a population
- Changes in allele frequencies ultimately lead to formation of new species

Modern theory of evolution

By the 1920's genetics were united with the evolutionary theory

Neodarwinism, or the synthetic theory of evolution
- Variation is a result of mutations
- Population genetics is the focus of evolutionary theory

Current debate on the speed of evolutionary processes

Evidence for evolution

Fossils
- Most in sedimentary rock
- Others frozen, mummified, in amber
- Fossil record is biased, and incomplete

- Conditions conducive for fossilization
    - Aquatic organisms become covered by sediments
    - Terrestrial organisms often decay quickly after death, or are destroyed by scavengers
    - Hard parts make the best fossils
- Older fossils are in deeper strata
- Radioactive dating allows us to exactly determine the age

Comparative anatomy
- Homologous structures indicate evolutionary ties
    - Example: wing of bird and wing of bat
- Analogous structures indicate similar evolutionary trends
    - Evolve to meet similar needs, but aren't related
    - Convergent evolution
    - Example: wings of insects and wing of birds
- Vestigial organs are remnants of organs that were previously used
    - Humans have 100+ vestigial structures
    - Example: pythons have vestigial hind leg bones buried in muscle

Biogeography
- Organisms evolve at the center of origin, disperse from there
- Because Africa and S. America were joined, their flora and fauna are rather similar.
- Organisms in N. America more closely resemble those in Europe because N. America and Eurasia were also joined

Resemblance of embryos
- Evolution doesn't act much on embryos; characteristics are conservative
- Embryos give clues as to evolutionary relationships
- Embryos of echinoderms are similar to chordates, rather than to other invertebrates; indicates a common ancestor
- Embryos of all vertebrates have similar characteristics

Biochemical comparisons
- Nearly all organisms rely on the same genetic code
- Proteins in related organisms have similar amino acid sequences
- Similarity of DNA indicates evolutionary ties

Research and Discussion Topics

• Investigate the development of the multidrug resistant strains of tuberculosis. What has caused the recent evolution of these strains of bacteria? Discuss the prevalence of tuberculosis in the US, and worldwide.

Suggested Readings

Dobzhansky, T. "Nothing in biology makes sense except in the light of evolution." *American Biology Teacher*. March 1973. 125-129. The classic paper on the central importance of evolution.

Gould, S.J. "Fall in the house of Ussher." *Natural History*. November 1991. 12-20. A description of the Archbishop of Ussher's hypothesis that the earth was created on October 23, 4004 b.c. at midday.

Perry, R.T. "Using different examples of natural selection when teaching biology." *American Biology Teacher*. April 1993. 55(4): 241-244. A description of some different examples of natural selection- different from the classic (tired) examples.

Caldwell, M. "Resurrection of a killer." *Discover*. December 1992. 59-64. A description of the current rise of tuberculosis.

Weiss, R. "On the track of 'killer' TB." *Science*. January 1992. 255: 148-150. A description of the drug-resistant strains of TB.

Mestel, R. "Ascent of the dog." *Discover*. October 1994. 90-98. A description of artificial selection which led to today's dog breeds.

Grant, P.R. "Natural selection and Darwin's finches." *Scientific American*. October 1991. 82-87. Long-term and short-term evolutionary processes on the Galapagos.

Chapter 18. Microevolution and Speciation

Chapter Overview

Populations are members of the same species which exist in a finite space and time frame. Populations exhibit variation, seen in the variety of alleles in the gene pool. Changes in the frequencies of alleles are known as microevolution. The Hardy-Weinberg principle is a mathematical model which describes the statistical frequencies of phenotypes and genotypes in a population which is at genetic equilibrium. The various assumptions of this model allow us to identify the factors which result in evolution.

Mutation is the raw material for variation in populations, but mutations are not directed, and may result in lessened fitness. Genetic drift is due to chance events which act on small populations. As a result of genetic drift, populations may change, again in a random fashion. The founder effect is a result of the small size of the original population, and bottlenecks are a result of a dramatic decline in a population. Both result in changes in allelic frequencies. Migration tends to decrease the amount of genetic variation between populations, and in that sense counteracts the other mechanisms mentioned previously.

Natural selection is the most important mechanism of evolution, and is directed towards increasing the number of favorable alleles in the population. Natural selection is a result of differential fitness, which results in differential reproduction of organisms within the population. Over time, the favorable alleles have a higher frequency within the population.

Three main mechanisms of selection exist. Stabilizing selection favors the intermediate phenotypes, directional selection favors one of the extreme phenotypes, and disruptive selection favors both extreme phenotypes. Selection is tied to variation within populations. This variation is enhanced in sexually reproducing organisms by independent assortment and crossing over during meiosis. Genetic polymorphism is therefore adaptive, and heterozygote advantage promotes this variation.

The biological species concept defines a species as a group of reproductively isolated organisms. Reproductive isolating mechanisms include prezygotic mechanisms (temporal, behavioral and mechanical) and postzygotic mechanisms (hybrid inviability or sterility). Speciation is the process of developing reproductive isolation. Allopatric speciation occurs when one subgroup of a population becomes geographically isolated, and this mode is particularly common in the formation of animal species. Sympatric speciation occurs without geographic speciation, and is particularly important in the evolution of plants. Plants often become polyploid, and allopolyploids are reproductively isolated from both parental species. This mechanism may explain the diversity of the angiosperms.

Lecture Outline

Variation in populations

- Population: a group of one species of organism existing in a specific place and time
- All of the genes within a population comprise the gene pool
  - Individuals can only have two alleles for each gene
  - Populations have many alleles for a particular gene
- Small changes in allelic frequencies within a population= microevolution

Hardy-Weinberg principle

- Mathematical model to describe frequencies of alleles and genotypes
- Defines the condition of genetic equilibrium (= no evolution)
- Conditions of equilibrium
  - Random mating; no selection of mates based on genotype
  - No mutations of alleles under consideration
  - Large population size, to decrease effect of chance
  - No migration; no exchange of genes with other populations
  - No natural selection
- This is the importance of HW equilibrium
- Evolution is deviation from HW equilibrium

Factors resulting in evolution

- Mutation
  - A permanent change in DNA
  - Source of new alleles
  - Are not necessarily positive
  - Contributes to the genetic variability of the population
- Genetic drift
  - In small populations, chance causes changes in allelic frequencies
  - Tends, however, to decrease diversity within the population
  - Founder effect is a type of genetic drift
  - Genetic bottlenecks may cause genetic drift
    - Seen recently in N. elephant seal and cheetah populations
- Migration of individuals and their genes
  - Reduces genetic variation between populations
  - Counteracts natural selection and genetic drift
- Natural selection
  - Darwin's mechanism of evolution
  - Variation results in some organisms surviving and reproducing
    - Favorable phenotypes have a selective advantage
  - Results in the accumulation of favorable phenotypes

Mechanisms of speciation

- Stabilizing selection
  - Somewhat "antievolutionary"- results in decreased variation

- Selection favors the intermediate phenotype
    - Bell-shaped curve of variation narrows
- Example: human birth weight

Directional selection

- Selection favors either extreme phenotype
    - Bell-shaped curve shifts to either extreme
- Example: peppered moths in England

Disruptive selection

- Selection favors the extreme phenotypes
- Splits into two forms, ultimately two species
    - Bell-shaped curve splits into two
- May act to promote sexual dimorphism
- Example: Galapagos finches

<u>Sources of variation</u>

- Variation is necessary for evolution
- Sexual reproduction is the source of variation
    - Independent assortment, crossing over both increase variety
- Genetic polymorphism
    - Presence of two or more alleles in a population for one trait
    - Gel electrophoresis can detect subtle differences
    - 25% of vertebrate genes are polymorphic
- Heterozygote advantage aids in maintaining variation
    - Heterozygotes have higher fitness than either homozygote
    - Example: sickle-cell anemia and malaria
- Neutral mutations do not change the fitness of the individual

<u>The species concept</u>

- A species is a reproductively isolated group; the gene pools do not mix
- Problems:
    - Only applies to sexually reproducing organisms
    - Tough to apply to extinct organisms
    - "Fuzzy" boundary between two groups that are in the process of speciation
    - Some organisms which are of different species may interbreed in zoos
- Reproductive isolating mechanisms
    - Prezygotic- act before a zygote is formed
        - Temporal isolation (seasonal isolation)
            - Have nonoverlapping reproductive periods
        - Behavioral isolation (sexual isolation)
            - Have species-specific courtship behaviors
        - Mechanical isolation
            - Reproductive parts are incompatible
    - Postzygotic isolating mechanisms
        - Nonviable zygote
        - Hybrid sterility, or hybrids exhibit behavioral isolation

Speciation

- Speciation occurs when a segment of the population becomes reproductively isolated from the rest of the population
- Allopatric speciation
  - A segment of the population becomes geographically isolated
  - Responsible for the evolution of most species of animals
  - Geographic isolation due to changes in the topography, or migration
  - May occur very rapidly
- Sympatric speciation
  - A segment becomes reproductively isolated while not geographically isolated
  - More common in plants
  - Common in plants which double the chromosome number
  - Polyploidy is a major factor in plant evolution
    - Allopolyploidy: polyploidy which occurs during sexual reproduction between two different species
    - Allopolyploids then cross fertilize with other similar plants, or self fertilize
    - 50% of all angiosperms are allopolyploids
    - Rapid mechanism for evolution
    - May have been the cause for the rapid evolution of angiosperms in the beginning of the Cenozoic

Research and Discussion Topics

• Describe the phenomena of the melanistic moths in England and industrial melanism. Note the directional selection towards darker forms during the Industrial Revolution, and the more recent shift back to lighter forms as air quality has improved.

• Other examples of heterozygote advantage include cystic fibrosis and cholera, tuberculosis and Tay Sachs disease. Research these conditions, and describe the cause for the relative high frequency of these genes which cause serious disease conditions or death.

• Discuss the evolution of bread wheat, a polyploid. It is very interesting to follow the various intermediate steps which ultimately produced the commercial wheat plant.

Suggested Readings

Bardell, D. "Some ancient-Greek ideas on evolution." *American Biology Teacher*. April 1994 56 (4): 198-200. An interesting description of the very beginnings of evolutionary thought.

Hammersmith, R.L. and T.R. Mertens. "Teaching the concept of genetic drift using a simulation." *American Biology Teacher*. Nov/Dec. 1990. 52 (8): 497-199. A lab designed to model genetic drift.

Mertens, T.R. "Introducing students to population genetics and the Hardy-Weinberg principle." *American Biology Teacher*. February 1992. 54 (2): 103-108. Advice on teaching students this perennially difficult subject.

Chapter 19. Macroevolution and the History of Life

Chapter Overview

Macroevolution refers to large-scale evolutionary changes. These changes may be due to novel uses for preexisting structures, regulatory changes during development or changes in timing of developmental events. These evolutionary changes may result in adaptive radiation. The mammalian adaptive radiation which occurred after the extinction of the dinosaurs is an excellent example. Extinctions are also tied to macroevolution. Background extinctions are constant and low-level, but mass extinctions, which coincide with the end of a geologic eras result in the loss of great numbers of taxa, and are typically followed by adaptive radiation of the surviving groups.

The pace of evolution is currently debated. Proponents of punctuated equilibrium believe that evolutionary time was marked by long periods of stasis, followed by rapid spurts of evolutionary change. Gradualists point out that the fossil record is biased and incomplete, and believe that evolution occurs slowly.

The early earth was characterized by a reducing atmosphere, dissolved mineral ions in the ocean, and large inputs of energy. Experiments have shown that, given time, and these environmental conditions, organic compounds form spontaneously. The first cells therefore arose from nonliving aggregations, called protobionts, which ultimately gained the ability to divide, concentrate organic molecules and carry out metabolism.

The first living organisms were aerobic heterotrophs, followed by phototrophs, which ultimately produced free oxygen, alloweing the evolution of aerobes. This oxygen also resulted in the formation of the ozone layer, which shielded these organisms from UV. Eukaryotes appeared about 2 million years ago, and are believed to have arisen from a mutualistic relationship between a host cell, aerobic bacteria, which ultimately became mitochondria, and aerobic phototrophs, which ultimately became chloroplasts.

Five eras are defined: the first two, the Archean and Proterozoic were characterized by massive topographic changes, and the dominance of prokaryotes and simple eukaryotes. The Paleozoic Era began with the Cambrian "Explosion," during which most modern phyla evolved. The Paleozoic Era was marked by a succession of firsts; first vertebrates, first land plants, first insects, etc. At the end of the Paleozoic Era, mass extinctions accompanied great topographic changes. The dinosaurs and other reptiles dominated in the Mesozoic Era, and after their extinction, the Cenozoic Era was marked by the adaptive radiation of angiosperms, mammals, insects and birds. Two novel habitats, expansive grasslands and angiosperm forests evolved during the Cenozoic. Further changes appeared as a result of the ice ages, including land bridges.

Lecture Outline

Macroevolution

- Large phenotypic changes occurring over relatively long periods of geologic time
- Changes in preexisting structures
  - Example: lungs evolved from the swim bladders of fish
  - Often due to regulatory changes in development
  - Also due to changes in the timing of development
- Adaptive radiation when an ancestral organism evolves to fill various niches (adaptive zones)
- Example: mammalian adaptive radiation after the Mesozoic Era

Extinctions

- Permanent loss of species
- Eventual fate of all species
- Has a positive evolutionary effect; opens up ecological niches
- Background extinctions
  - Continuous extinctions of species
- Mass extinctions
  - Occurred 4 or 5 times in history
  - May have been triggered by large environmental catastrophes
  - Often followed by massive adaptive radiations of survivors
  - We are currently in a biologically (human)-caused mass extinction
- Progress of evolution may be rapid or slow
  - Fossil record often lacks intermediate stages in evolution
  - Punctuated equilibrium states that evolution occurs in spurts
    - Long periods of stasis followed by rapid radiation
  - Gradualistic view holds that the fossil record is simply missing the intermediate forms
- The pace of evolution probably follows both, at different times

Early life on earth

- First life evolved from nonliving molecules
- Conditions on early earth
  - Life evolved in the oceans
  - Atmosphere was a strongly reducing atmosphere (no free O_2)
  - Atmosphere with carbon dioxide, water vapor, carbon monoxide, hydrogen and nitrogen gas
  - Strong UV light probably broke down most other molecules like methane, ammonia, and hydrogen sulfide
  - The sources of energy were storms, volcanoes and solar radiation
  - Chemicals dissolved in water were ions and the atmospheric gases
  - Over long periods of time, organic molecules were formed by chance
- Oparin and Haldane first theorized this process

Miller and Urey tested it
Found organic molecules formed by chance
More recent tests have shown amino acids and nucleotide bases may be formed in these conditions
Older hypotheses- life evolved in the primordial soup
Newer ideas- clay surfaces polymerized first organic molecules
Clay has catalytic charged sites
Protobionts- first aggregations of organic polymers
Eventually protobionts gained ability to divide, concentrate organic molecules inside, show signs of metabolism
First cells were prokaryotic
Fossils approximately 3.5 billion years old
Stromatolites (cyanobacterial mats), found in Australia, Lake Superior
First cells were heterotrophs, used anaerobic fermentation pathways
Phototrophs evolved next; advantageous because these cells did not have to compete for the preformed organic molecules
First phototrophs relied on H_2S
Cyanobacteria (blue green algae) were the first aerobic phototrophs
Produced first free oxygen
Spurred the evolution of aerobic heterotrophs
Aerobic respiratory pathways were "added" on to glycolysis
First oxygen also reacted and formed first ozone
Protected earth from UV

Evolution of Eukaryotes

Eukaryotes appeared approximately 2 million years ago
Eukaryotes evolved from prokaryotes
Endosymbiont theory
Organelles such as mitochondria and chloroplasts evolved from symbiotic prokaryotes
Chloroplasts evolved from photosynthetic bacteria
Mitochondria evolved from aerobic heterotrophic bacteria
May have originally been ingested, ultimately became mutualists
Evidence:
Mitochondria and chloroplasts each have their own DNA
Mitochondria and chloroplasts are both double-membrane bound
Also have own ribosomes, and can carry out protein synthesis
Does not explain the origin of the nucleus

Fossil evidence of life on earth

Five major rock strata, subdivided into minor strata
Five major eras
Archean, Proterozoic, Paleozoic, Mesozoic, and Cenozoic
Eras subdivided into periods, subdivided into epochs

Archean Era began 3.5 billion years ago, after crust formation
- Few fossils (cyanobacteria), deeply buried
- Seen in Grand Canyon, few other places
- Lasted 2 billion years
- Widespread crust movement, deformation

Proterozoic Era- 1.5-0.5 billion years ago
- Characterized by deposition of sediment
- Fossils of bacteria, fungi, protists, multicellular algae, simple animals
- No vertebrate fossils

Paleozoic Era- 570-248 million years ago
- Cambrian Period
  - Cambrian Explosion- rapid appearance of most modern phyla
  - Marine fossils such as clams, trilobites, sponges, cnidarians
- Ordovician Period
  - Marine fossils including shelled cephalopods, first jawless vertebrates
- Silurian Period
  - First terrestrial animals (arachnids)
  - First plants (fern-like), lived in wet areas
    - Had vascular tissue
- Devonian Period
  - The Age of Fishes
  - First amphibians
  - Diversification of vascular plants
- Carboniferous Period
  - Swamp forests (today's coal deposits)
  - First reptiles
  - Winged insects
- Permian Period
  - Seed plants dominated
  - Massive topographic changes
  - Mass extinctions

Mesozoic Era- 248-65 million years ago
- "The Age of Reptiles"
- Dinosaurs included the largest animals that ever lived
- Current ideas about dinosaurs include the possibility of social behavior and endothermy
- Evolution of modern insect orders
- Triassic Period
  - First mammals
  - Gymnosperms dominated
- Jurassic Period
  - First birds, <u>Archaeopteryx</u>
- Cretaceous Period
  - Angiosperms dominated by the end
  - Dinosaurs and many other plants and animals became extinct at the end

Cenozoic Era
"The Age of Mammals, Birds, Insects and Flowering Plants"
Tertiary Period
Appearance of grasslands and forests
Increase in size of mammals, appearance of large grazers
Mammalian brain size increased, feet and teeth highly adapted
Quaternary Period
4 ice ages
Ice ages caused land bridges; exchange of organisms between N. America and Asia
Many large mammals became extinct due to human hunting

Research and Discussion Topics

• Various experiments have shown that in the early life, organic molecules were formed spontaneously from inorganic. Why don't we see this "spontaneous generation" today?

• Describe the formation of stromatolites. What organisms form stromatolites? Where are they currently found?

Teaching Suggestions

• Students may have a hard time memorizing the time periods. One nemonic for the periods of the Paleozoic and the Mesozoic is: "Camels Often Sit Down Carefully, Perhaps Their Joints Creak". Encourage students to come up with an even more creative slogan!

Suggested Readings

Horgan, J. "In the beginning..." *Scientific American*. February 1991. 116-125. A description of the early earth; the formation of organic molecules, RNA as a catalyst, stromatolites etc.
and
Radetsky, P. "How did life start?" *Discover*. November 1992. 74-82.

Allegre, C.J. and S.H. Schneider. "The evolution of the earth." *Scientific American*. October 1994. 66-75. Two articles describing the beginnings of life on earth.

Orgel, L.E. "The origin of life on the earth." *Scientific American*. October 1994. 76-83. The involvement of catalytic RNA.

Gould, S.J. "The evolution of life on the earth." *Scientific American*. October 1994. 84-91. Evolution, complexity and extinctions.

Wellnhofer, P. "*Archaeopteryx*." *Scientific American*. May 1990. 70-77

Alters, B.J. and W.F. McComas. "Punctuated equilibrium: the missing link in evolution education." *American Biology Teacher*. September 1994. 56 (6) 334-340.
and
McComas, W.F. and B.J. Alters. "Modeling modes of evolution: comparing phyletic gradualism and punctuated equilibrium." *American Biology Teacher*. September 1994. 56 (6) 354-360. Two articles which introduce a description of Eldredge and Gould's ideas about punctuated equilibrium, compared to Darwinian phyletic gradualism, followed by a lab which models a punctuational scheme.

Gore, R. "Dinosaurs." *National Geographic*. January 1993. 8-52. A classic National Geographic treatment; lots of illustrations, new theories about dinosaurs.

Morell, V. "Announcing the birth of a heresy." *Discover*. March 1987. 26-50. A discussion of Jack Horner, Robert Bakker and others; new ideas about dinosaurs.

Chapter 20. The Classification of Organisms

Chapter Overview

The modern system of biological nomenclature originated with Carolus Linnaeus, who devised the binomial system; every extant and extinct species has a unique genus and species name. Taxa are taxonomic groupings, and range from broad groupings; the 5 kingdoms, to the species and subspecies levels. Species have a biological basis (reproductive isolation from other species), but other levels represent evolutionary ties.

Systematics is based on phylogeny (evolutionary history), and systematists arrange taxa typically in monophyletic groupings. Monophyletic taxa share common ancestors; polyphyletic taxa are rather artificial and include organisms with similar characteristics, but not common ancestors. Systematists analyze homologous structures, compare ancestral and derived structures, as well as molecular similarities.

Phenetics is a numerical system which evaluates the number of shared characteristics, whereas cladistics use phylogenetic evidence to classify organisms. Classical evolutionary taxonomy uses both evolutionary branching and divergence to classify organisms.

Lecture Outline

Binomial nomenclature

- Linnaeus, Swedish botanist developed system of binomial nomenclature
- Scientific name consists of genus and the specific epithet
- Correct format: *Perca flavescens*, the yellow perch
- Typically derived from Greek and Latin names
    - Descriptive, named for people, places

Taxonomic classification

- Kingdom is the most broad classification
- Taxa are any level of taxonomic classification
- Species are organisms that can interbreed and are reproductively isolated from members of other species
    - The species is the only biologically based classification, although other groupings reflect evolutionary relationships
    - Subspecies are smaller classifications than species
    - For bacteria, the term strain is used
- Species are grouped into genera, genera into family, families into orders, orders into classes, classes into phyla (divisions for plants) and phyla into kingdoms
- Various other classifications are useful; superfamilies, subspecies, infraorders

Some scientists have suggested classification of the domain
    Three domains: archaebacteria, eubacteria, and eukaryotes
Lumpers versus splitters
Five phyla are typically recognized: Prokaryotae, Protista, Fungi, Plantae and Animalia

Systematics

- Classification based on evolutionary relationships
- Phylogeny= evolutionary history
    - Monophyletic taxa share a common ancestor
        - Share homologous structures
        - Share ancestral characters
    - Polyphyletic taxa are rather artificial, grouping taxa which do not share common ancestors
        - May show analogous structures
    - Derived characteristics evolved relatively recently
        - Show branching points among taxa
- Taxonomy is a dynamic science, always changing
    - Most recently discovered phyla, Loricifera, described in the 1980's
    - Molecular biology gives clues to taxonomists
        - Changes in genes, proteins can be used as molecular clocks
        - Recent evidence shows that fungi may be more closely related to animals than plants based on ribosomal RNA
- Phenetic system (numerical taxonomy) is based on the degree of similarity between taxa
    - Characters are rated as present (+) or absent (-), then counted
    - Does not necessarily indicate evolutionary relationships
- Cladistic system focuses on evolution as basis for classification
    - Use cladograms which show evolutionary "tree"
- Classical evolutionary taxonomy considers both phylogenetic branching and extent of divergence

Research and Discussion Topics

• Compare and assess the various methods of determining the total number of species on the earth. Refer to May, R.M. "How many species inhabit the earth?" *Scientific American*. October 1992. 267 (4): 42-48.

• Investigate the changes in taxonomy since Linnaeus. Note major changes, such as the later recognition that sponges are not plants (Linnaeus thought they were), the changing classification of barnacles (Linnaeus thought they were bivalves), and the breakdown of Linnaeus' phylum Vermes (worms).

• Approximately how many animal phyla are currently recognized? Which are the most speciose? Which are the least speciose?

• Draw a cladogram of the primates. Include *Homo sapiens*, and the other extinct species of the genus *Homo*.

Teaching Suggestions

• A nemonic for remembering the taxonomic heirarchy is this: Keep pots clean or family gets sick.

Suggested Readings

Jukofsky, D. and C. Wille. "They're our rainforests too." *National Wildlife*. April/May 1993. 18-37. A description of rainforest products in our lifes.

Wilson, E.O. "Rain forest canopy; the high frontier." *National Geographic*. December 1991. 78-107. Research in the tropics.

Blaustein, A.R. "Amphibians in a bad light." *Natural History*. October 1994. 32-39. Possible causes for the global amphibian decline.

Terborgh, J. "Why American songbirds are vanishing." *Scientific American*. May 1992. 98-104. Declines in songbirds, particularly the neotropical migrants.

Anonymous. "Meet the new bug on the block." *Science '83*. December 1983. 6. A description of the loriciferans, the most recently described phylum.

Chapter 21. Microorganisms: Viruses, Bacteria, and Protists

Chapter Overview

Viruses are nonliving particles which rely on living cells for their reproduction. Viruses are capsids surrounding genetic material (DNA or RNA). Viruses may have "escaped" from bacteria, plant or animal cells, and are rather specific as to their host. Bacteriophages infect bacteria. Lytic bacteriophages attach, penetrate, and inject their genetic material into the host cell. It uses the bacteria's ribosomes and enzymes to replicate their genetic material, proteins etc. to make new viruses, which are released from the bacterium when it lyses. Temperate viruses use the host cell's DNA, but do not lyse the cell.

Various diseases of plants and animals are caused by viruses. An unusual group of RNA viruses which infect animals are the retroviruses, which includes HIV. Plant viruses are spread between plants by plant-feeding insects like aphids, and spread between cells via plasmodesmata.

Bacteria, members of Kingdom Prokaryotae, lack membrane-bound organelles, and have their DNA arranged in one main ring, and accessory rings known as plasmids. Bacteria can produce capsules or form spores to avoid harsh environmental conditions. Bacteria may be saprobic (heterotrophs which gain nutrition from dead organic material), or autotrophic (phototrophic or chemotrophic). Bacteria are ecologically important in nitrogen fixation, as well as their role as decomposers.

The archaebacteria are unusual bacteria which are adapted to very harsh environments, and include halophiles, methanogens and thermophiles. The eubacteria are a very diverse group, but have only three main morphological forms; the bacillus, spirillum, and coccus. They may be classified as gram-positive or gram-negative bacteria, or as a mycoplasma, an odd group which lacks a cell wall.

Gram-negative bacteria have thin cell walls, and do not stain with a gram stain. Different gram-negative bacteria cause diseases, or fix nitrogen; others are enteric decomposers and cyanobacteria. Gram-positive bacteria have thick walls of peptidoglycan, stain darkly with a gram stain, and include the lactic acid bacteria, streptococci, staphylococci, and the clostridia.

Protists are a diverse group of eukaryotes which may be autotrophic or heterotrophic; most are aquatic, but some inhabit moist terrestrial habitats, or the bodies of other organisms. The slime molds and water molds resemble fungi, but some produce flagellated cells (fungi do not have flagella). They have very unusual life cycles: the slime molds have alternating body forms within their life cycle and reproduce with spores. One water mold infects potatoes and resulted in the Irish potato famine in the 1840's.

Algae are primarily autotrophic protists, and include dinoflagellates (ecological importance: as zooxanthellae, causing red tides, paralytic shellfish poisoning), diatoms (ecological importance: major producers in cool marine ecosystems), euglenoids (ecological importance: indicators of organic pollution), green algae (evolutionary importance: ancestor of higher plants) and the red and brown algae (ecological importance: macroalgae in marine environments).

The protozoa are the animal-like protists, and include amoebas and foraminiferans which move by extensions of the cytoplasm, flagellates which use flagella to pull the cell through the medium, ciliates which use cilia to move, and sporozoans which are nonmotile. The flagellate *Trypanosoma* causes African sleeping sickness, and the sporozoan *Plasmodium* causes malaria.

Lecture Outline

Viruses

- Structure
  - Nucleic acid core and protein coat (capsid)
  - May have an outer membrane
  - Contain DNA or RNA
  - Shape of capsid: rod shaped, and/or polyhedral
- Not truly living as they are not composed of true cells, cannot reproduce or carry out metabolic activities on their own
- Origin: may be bits of DNA or RNA that "escaped" from the parental cell, which may have been plant, animal, or bacterial cells
  - This theory explains specificity of viruses to host cells
- Bacteriophages
  - Infect bacteria
  - Typically DNA viruses in polyhedral head
  - May have a tail with attachment fibers
  - Easily cultured in the laboratory
    - Often used in biotechnology
  - Lytic bacteriophages
    - Destroy the host cell
    - Stages of infection:
      - Attachment
      - Penetration- nucleic acid injected into bacterium
        - Capsid remains attached to bacterial cell wall
      - Replication of new bacteriophage molecules using bacteria's ribosomes, enzymes, etc.
      - Assembly of new bacteriophages
      - Release of bacteriophages by bacterial lysis
  - Temperate bacteriophages
    - Do not destroy the host cell
    - Integrate their DNA into host DNA, replicate together

May, however, change to a lytic cycle
Example: bacteria which causes diphtheria; the strain which causes the disease harbors a temperate bacteriophage

Viruses infecting animal cells
- Causes diseases including chicken pox, genital herpes, mumps, rubella, measles, rabies, warts, mononucleosis, influenza, hepatitis
- Animal cells may engulf viruses by endocytosis
  - Viruses then replicate within the animal cells
- Retroviruses, which include HIV, use reverse transcriptase to transcribe RNA into DNA, then use the DNA to make copies of viral RNA
- Viruses leave the host animal cell by lysis or exocytosis

Viruses infecting plant cells
- Spread between plants by plant-feeding insects like aphids
- May also be spread by seeds, or asexual propagation of plants
- Pass between plant cells via plasmodesmata
- Most are RNA viruses; attach to plant ribosomes, then translated
- Symptoms of viral infection: spots on leaves, lower yields
- "Cures" for plant viral diseases are not well known

Bacteria

Characteristics:
- Single celled, contain ribosomes, lack membrane-bound organelles
- DNA is not surrounded by a nuclear membrane, is circular
  - Accessory DNA rings are known as plasmids
- Have a cell wall exterior to the cell membrane
  - Typically composed of peptidoglycan; sugars linked with peptides
  - May have a capsule exterior to the cell wall
    - Aids in protection against desiccation
    - Adds to disease-causing ability
- May have flagella; very different from eukaryotic flagella
  - Pushes the cell through the medium (eukaryotic flagella pull)
- May have hairlike structures which aid in attachment to a substrate
- May have tube-like pilli to transmit DNA between bacteria

Bacterial metabolism
- Heterotrophic bacteria
  - Most are saprobes; gain nourishment from dead and decomposing matter
  - Important as decomposers; recycle organic matter
- Autotrophic bacteria are either phototrophs or chemotrophs
  - Phototrophs are photosynthetic
  - Chemotrophs oxidize inorganic molecules as an energy source

Bacterial reproduction
- Bacteria typically reproduce by binary fission
- DNA is replicated, new cell membrane and wall forms between the two rings of DNA

Very rapid; as fast as every 20 minutes
In reality, this potential is limited by lack of food, accumulation of wastes
Nitrogen fixation is critical to life on earth
Bacteria which have this ability take atmospheric nitrogen and convert it to organic forms, which are then available to plants

<u>Archaebacteria</u>

Ancestral prokaryotes split into 2 groups early in the history of life
Tend to live in environments of harsh environmental conditions
Lack peptidoglycan in cell walls
Other biochemical differences from eubacteria
Actually have some characteristics in common with eukaryotes
Halophiles live in very salty environments, some are photosynthetic
Methanogens (most common type) produce methane
Inhabit digestive tract of humans and other animals
Thermophiles live in very hot and often acidic environments
Found in Yellowstone Park, deep ocean vents

<u>Eubacteria</u>

A very diverse group, yet similar morphologies
Three main shapes: bacillus, coccus, and spirillum
Gram staining
Gram-positive bacteria stain violet
Thick walls of peptidoglycan
Penicillin works most effectively against gram-positive bacteria
Gram-negative bacteria do not absorb the stain
Two layered wall; thin inner peptidoglycan wall, and an outer membrane
Mycoplasmas are tiny bacteria without a cell wall
Live in soil, sewage, some are parasites
Typically not pathogenic
Gram-negative bacteria
Pathogenic bacteria including gonorrhea, syphilis, typhus, Rocky Mountain spotted fever, bacterial meningitis
Nitrogen-fixing bacteria
Fix nitrogen; convert N_2 to NH_3
Some are free-living, others live in root nodules
Enterobacteria are decomposers, and include *E. coli*
<u>E. coli</u> normally inhabit our GI tract; some strains can cause diarrhea, even death
Cyanobacteria (blue-green algae) are aquatic, terrestrial in moist areas
Photosynthetic autotrophs
Many also fix nitrogen
Gram-positive bacteria
Lactic acid bacteria produce lactic acid from fermentation of lactose
Used commercially to make yogurt, pickles and sauerkraut

- Streptococci cause "strep throat," tooth decay, pneumonia, scarlet fever, rheumatic fever
- Staphylococci normally live in the nose and skin
    - Cause disease when the immune system of the host is lowered
    - May cause boils, skin infections, food poisoning, toxic shock syndrome
- Clostridia are anaerobes; cause gas gangrene, tetanus, botulism

<u>Protists</u>

- Diverse group; some plant-like, others animal-like
- Most are unicellular, some colonial, coenocytic or multicellular
- Modes of nutrition:
    - Phototrophs include the algae
    - The heterotrophs include the water molds and the protozoans
- Habitats:
    - Most are aquatic, and make up part of the plankton
    - Others live in moist terrestrial habitats, or are parasitic
- Slime molds and water molds
    - Resemble fungi as they are heterotrophic, and body form is threadlike
    - Have flagellated cells (fungi do not)
    - Plasmodial slime molds (Phylum Myxomycota)
        - Feeding stage is the multinucleate plasmodium
        - Streams over decaying matter
        - Reproductive stage includes stalked structures with sporangia which produce haploid spores
        - Spores grow into another plasmodium
    - Cellular slime molds (Phylum Acrasiomycota)
        - Feeding stage is an individual ameboid cell
        - Reproduction occurs when the cells aggregate as a pseudoplasmodium.
        - Produces a fruiting body which has spores
        - Spores grow into the feeding individuals
    - Water molds (Phylum Oomycota)
        - Resemble a fungal mycelium; have hyphae which are coenocytic
        - Have flagellated reproductive cells
        - Caused the Irish potato famine in the 1840's
- Algae
    - Primarily autotrophs
        - Have chlorophyll *a* and *b*, as well as others, such as carotenoids
    - Dinoflagellates (Phylum Dinoflagellata)
        - Unicellular, with 2 flagella, shell of cellulose plates
        - Many are symbiotic in jellyfish, coral, molluscs, known as zooxanthellae
            - Critical to coral reefs
        - May form water blooms, which may form red tides
        - Paralytic shellfish poisoning (PSP) results from eating filter feeders which concentrate the dinoflagellate toxins

- Diatoms (Phylum Bacillariophyta)
  - Unicells, some are colonial
  - Siliceous shells of two parts, distinctive pattern of lines and dots
  - Freshwater and marine, particularly abundant in polar waters
  - Diatomaceous earth is a fossil deposit of diatom shells
- Euglenoids (Phylum Euglenophyta)
  - Unicellular flagellates, many are photosynthetic
  - Two flagella (only one is emergent and functional)
  - Flexible and unusual outer covering (not a cell wall)
  - An indicator of organic pollution
- Green algae (Phylum Chlorophyta)
  - Various body forms; unicells, multicellular tubes, sheets, balls
  - Primarily freshwater, others are marine, some terrestrial
  - Very similar to higher plants, thought to share common ancestor
- Red algae (Phylum Rhodophyta)
  - Typically multicellular, form tufted or leaflike body
    - Coralline reds are very tough and resistant to herbivory
      - Important in coral reefs
  - Attach to substrate by holdfast (not a true root)
  - Commercial uses- agar, carageenan, eaten by humans as well
  - Dominate in warm oceans
- Brown algae (Phylum Phaeophyta)
  - Largest protists; all are multicellular
  - Kelp have leaflike blades, stemlike stipes, and rootlike holdfasts
    - Important as a source of algin, also eaten by humans
  - Dominate in colder oceans
  - Important in structuring shallow offshore habitats
    - Provide both food and habitat for a variety of organisms

Protozoans; the animal-like protists

- Amoebas (Phylum Rhizopoda)
  - Unicellular, changeable body shape
  - Move, ingest food by forming pseudopodia
  - *Entamoeba* causes amoebic dysentery in humans
- Foraminiferans (Phylum Foraminifera)
  - Marine unicells which produce chalky tests
  - Feed by extending pseudopods
  - Dead foram deposits form chalk; make up the Great White Cliffs of Dover
- Flagellates (Phylum Zoomastigina)
  - Unicells which move with flagella
  - Eukaryotic flagella pull the organism through the water
  - Are heterotrophs
  - Includes *Trypanosoma*
- Ciliates (Phylum Ciliophora)
  - Unicells which move with cilia
  - May have trichomes which discharge filaments to trap prey

Includes *Paramecium*
Most are motile, some are stalked like*Vorticella*
Sporozoa (Phylum Apicomplexa)
Spore forming parasites
Spores are resistant structures; often are the infective agent
Lack structures for locomotion
Includes *Plasmodium*, which causes malaria

<u>Evolution of eukaryotes</u>
Protists were the first eukaryotes
Evolved over 2 billion years ago
Organelles (mitochondria and chloroplasts) evolved as endosymbionts
Multicellularity evolved several times; in the green, red and brown algae
Higher eukaryotes may have evolved from green algae, such as *Chlamydomonas* and its relatives like *Volvox*

Research and Discussion Topics

• Discuss the pros and cons of dividing Kingdom Prokaryotae into 2 groups, the archaebacteria and the eubacteria. Should these be two kingdoms, or a larger group, the domain?

• Some current research has suggested that red tides are increasing in prevalence. Some work done in the last few years in North Carolina indicates that some dinoflagellates produce toxins to kill fish, which the dinoflagellates then consume! (Huyghe, P. "Algae." *Discover*. April 1993) Refer to the Scientific American article for further information and references. Anderson, D.M. "Red Tides." *Scientific American*. August 1994. 271 (2): 62-38.

• Investigate the research which indicates that green algae may have been the ancestors of higher plants. What cellular evidence is there for this theory?

Teaching Suggestions

• I find that few subjects (except sex; see chapter 42) hold student's interest more than diseases. I always spend time talking about bacteria, viruses and the diseases they cause. We discuss ways of preventing and treating them, related to whether they are caused by bacteria or viruses. We discuss some of the diseases that were of importance in history, like smallpox and typhus, and some of recent concern, like AIDS, TB and the hepatitis group. I also spend some time talking about STD's, and feel that this is a bit of a "public service message,", as I discuss modes of transmission, symptoms and treatment. Nothing like a picture of a chancre to get their attention!

• Discuss "food poisonings," including that caused by *Salmonella, E. coli*, botulism, listeriosis. Students find this very interesting

Suggested Readings

Canby, T.Y. "Bacteria." *National Geographic*. August 1993. 184 (2): 36-61. Bioengineering and bacteria.

Gillen, A.L. and R.P. Williams. "Dinner date with a microbe." *American Biology Teacher*. May 1993. 55 (5): 268-274.
and
Williams, R.P. and A.L. Gillen. "Microbe phobia and kitchen microbiology." *American Biology Teacher*. January 1991. 53 (1): 10-11. Two articles which discuss microorganisms which produce food products for us.

Brown, W.E. and R.P Williams. "Ignaz Semmelweis and the importance of washing your hands." *American Biology Teacher*. May 1990. 52 (5) 291-294. A fascinating account of the beginnings of antiseptics and a lab to investigate the effect of Chlorox on bacterial growth.

Wallis, C. "Viruses." *Time*. November 3, 1986. 66-74. Simplified description of viruses, replication, diseases.

Tollais, P. and M. Buendia. "Hepatitis B virus." *Scientific American*. April 1991. 116-123. Hepatitis B, cancers, vaccines.

Langone, J. "Emerging viruses." *Discover*. December 1990. 63-68. Overshadowed by AIDS, other viruses like Embola and yellow fever still haunt us.

Radetsky, P. "Forgotten but not gone." *Discover*. September 1989. 22-24. A description of polio today.

Caldwell, M. "Vigil for a doomed virus." *Discover*. Mary 1992. 50-53. A description of the execution of the smallpox virus.

Diamond, J. "Blood, genes and malaria." *Natural History*. February 1989. 8-18. A description of sickle-cell anemia and malaria.

Tennesen, M. "Kelp, keeping a forest afloat." *National Wildlife*. June/July 1992. 5-11. Changes in California's kelp forests.

Maxwell, C.D. "A seaweed buffet." *American Biology Teacher*. March 1991. 53 (3): 159-161. Description of edible algae; includes some recipes!

Kunzig, R. "Invisible garden." *Discover*. April 1990. 66-74. The importance of planktonic diatoms, dinoflagellates and coccolithophorids.

Moser, P.W. "It must have been something you ate." *Discover*. February 1987. 94-100. An interesting description of food poisonings.

Chapter 22. Fungal Life

Chapter Overview

The organisms classified in Kingdom Fungi are primarily terrestrial eukaryotes, characterized by unique cell walls, and lack of chloroplasts, so they are therefore heterotrophs (primarily saprobes). Their body plan, except for the unicellular yeasts, is filamentous; made up of hyphae which form a mycelium. The hyphal filaments may be coenocytic, or have septa, which have perforations. In either case, the cytoplasm streams readily within the hyphae. Fungi are classified by their mode of reproduction; most reproduce via spores, which are nonmotile, and are produced in a variety of different structures called fruiting bodies. In sexual reproduction, different mating types fuse, forming a hypha with two nuclei that may or may not fuse.

The zygomycetes are decomposers like the black bread mold, which sexually reproduce by the fusion of two different hyphae, forming a diploid nucleus. After germination, haploid spores are produced, ultimately forming a haploid mycelium. Ascomycetes, the sac fungi, are the most speciose group, including yeasts, molds and morels. Asexual spores, conidia, are produced on conidiophores. In sexual fertilization, the fruiting body is known as an ascocarp, and ascospores are formed within the asci in the ascocarp. The zygotes are the only diploid cells in the life cycle. In contrast, the basidiomycetes, or club fungi form a fruiting body, the basidiocarp (mushroom). The basidia, on which basidiospores form, is on the gills of the mushroom.

The lichens are dual "organisms," composed of an alga and a fungus. The alga does well "on it's own", but the fungus does not, and so the lichen has been described as a controlled parasitism of the alga by the fungus. Lichens can tolerate environmental extremes, but they are sensitive to sulfur dioxide (a component of acidic precipitation). Lichens typically reproduce asexually by fragmentation.

Fungi are ecologically important in their role as decomposers, as well as their part in mycorrhizae. They are also economically important, both in a negative (decomposition, cause a few diseases), and positive way. We use fungi (yeasts) to ferment beer and wine and to make leavened bread. Commercially, fungi are used to make blue and ripened cheese, and soy sauce, and mushrooms, morels and truffles are eaten by humans. Penicillin is an antibiotic derived from *Penicillium*.

Lecture Outline

<u>Characteristics of fungi</u>

- Eukaryotic
- Mostly terrestrial
- Cell walls of chitin (a polysaccharide polymer)
    - Very resistant to decay

- Lack chlorophyll and chloroplasts
    - Are heterotrophic, but do not ingest food; they absorb it
    - Are decomposers or parasites
- Grow best in dark, moist habitats (no competition there with plants)
    - Require moisture; if dried go dormant or produce spores
    - Can tolerate a wide variation in pH
    - Can tolerate a wide variety of osmotic pressures in their medium
    - Live over a wide variety of temperatures
    - Therefore can live in the back of your refrigerator!
- Body plans
    - Single celled fungi include yeasts
        - Reproduce asexually by budding or fission
        - Sexual reproduction by spore formation
    - Filamentous molds
        - Threads called hyphae form a mycelium
        - Some hyphae are coenocytic, others have septa between cells
            - Septa have pores to allow transfer of cytoplasm and organelles between cells
- Reproduction
    - Spores are nonmotile haploid reproductive cells
    - Since they are nonmotile, fungi produce structures to hold up the spore producing parts
        - Fruiting bodies; mushrooms are a familiar example
    - Spores are produced sexually or asexually
        - Fungal cells contain haploid nuclei; in sexual reproduction, haploid nuclei fuse forming a diploid zygote
        - In some, the nuclei fuse (monokaryotic)
        - In others, the nuclei remain distinct (dikaryotic)
    - Spores germinate and grow when on a nutritive substrate
        - Hyphae grow through the medium
        - The fungus digests and absorbs the substrate
        - Later it forms a fruiting body and spores

<u>**Fungal classification**</u>

- Classified on the structure of the fruiting bodies, and the sexual spores
- 3 sexually reproducing divisions are typically recognized
- One division (Deuteromycota) do not show sexual stages and are members are simply lumped together for convenience
- Division Zygomycota; the zygomycetes
    - Sexual spores are called zygospores, and remain dormant for long periods of time
    - Coenocytic hyphae
    - Are mostly decomposers
    - Example: *Rhizopus*, the common black bread mold
        - The sporangia are the evident black dots
    - Sexual reproduction occurs between + and - strains (not male and female); are different mating types

Hyphae of opposite mating types meet and fuse, forming a diploid zygote (only stage which is diploid)
- Meiosis occurs at germination
- The zygospore forms an aerial hypha with a terminal sporangium
- Mitosis within the sporangium forms haploid spores

Division Ascomycota includes the sac fungi
- Very speciose group
- Sexual spores are produced in sacs called asci
- Hyphae have perforate septa
- Includes most yeast, many molds, the cup fungi and the edible morels and truffles
- Plant diseases such as Dutch elm disease, chestnut blight, ergot and powdery mildew
- Asexual reproduction
  - Produce spores called conidia on specialized hyphae (conidiophores)
  - May self fertilize
- Sexual reproduction between different mating strains
  - Hyphae from 2 strains grow together, but nuclei do not fuse
  - This produces hyphae which are dikaryotic
  - The n+n hyphae form the fruiting body, the ascocarp
  - Asci are produced in the ascocarp
    - A diploid zygote is formed
    - The zygote undergoes meiosis to form 4 haploid nuclei
    - These ultimately form 8 haploid nuclei
    - Each form a haploid ascospore, which disperse in the air
- Yeast are unicellular ascomycetes
  - Yeast reproduce by budding, or by forming ascospores
  - Commercial use in making bread and beer and wine

Division Basidiomycota are the club fungi
- Also a large division; includes fungi and puff balls
- Includes plant parasites; rust and smut
- Hyphae are divided by perforate septa
- Sexual reproduction
  - Form basidia, which are similar to the asci of ascomycetes
    - Basidiospores form on the outside of the basidia
  - Cultivated mushrooms are the fruiting body of the fungus
    - The stalk and cap are the basidiocarp
    - Gills develop on the lower surface of the cap, bearing basidia on their surface
  - Basidiospores give rise to a primary mycelium, which are monokaryotic cells
  - Meets a hypha of a different mating type and fuses
    - Secondary mycelium is dikaryotic
  - Secondary hyphae form basidiocarps

On gills, dikaryotic nuclei fuse forming diploid zygotes
Zygotes are again the only diploid cells present
Meiosis immediately forms four haploid nuclei, the basidiospores
Basidiospores are released
Division Deuteromycota, the imperfect fungi
Never seen to have sexual stages
Most reproduce by conidia
Many appear to be related to ascomycetes; others to basidiomycetes
Lichens are symbiotic associations between a phototroph and a fungus
The autotroph is usually a green alga or cyanobacterium
The fungal partner is usually an ascomycete
Algae can be found as free-living species
When cultured separately, grow faster without the fungus
Fungi are typically only found within the lichen
When cultured, the lichen grows more slowly without the fungus
Originally described as mutualism
Now described as a controlled parasitism of the alga by the fungus
Lichens tolerate climatic extremes
Intolerant of pollution, particularly sulfur dioxide
Ecological importance: reindeer mosses are lichens; food for caribou
Commercial importance: produce dyes to make margarine yellow, dye litmus paper
Reproduce primarily asexually by fragmentation
The fungal component may reproduce by forming ascospores

Ecological and economic importance of fungi
As saprophytes; they decompose dead organic matter
May be mutualists
Mycorrhizae are the intimate association between fungi and plant roots
Seen in 90% of plant families
Benefits the fungus by decomposing organic material in the soil around the roots of the plant
Mutual transfers of nutrients between roots and fungi
Many gymnosperms are dependent on mycorrhizae
Decomposition causes economic harm to humans
Yeasts are used to make beer, wine and bread
Cheeses, such as the ripened and blue cheeses, are produced by the action of molds
Aspergillus produce soy sauce
Many fungi are edible, although the few poisonous ones are to be avoided
Various other mushrooms are hallucinogenic
Penicillin is produced by *Penicillium*; fungi also produce cyclosporine
Many plant diseases are caused by fungi; mildews, smuts and rusts

Fungi may cause superficial infections like ringworm and athlete's foot
Others cause common vaginal infections
Many fungal infections are seen in persons with depressed immune systems, such as AIDS patients.

Research and Discussion Topics

• It has been suggested that the Salem, Massachusetts witch trials were caused by ergot poisoning of women, who were believed to be "bewitched." Investigate the causes and symptoms of ergotism.

• Investigate the role of yeast in the production of beer and wine. What are the functions of the various components of beer; hops, malt, yeast, barley? How is wine made? Explain the difference between champagne and still wine. What does the term residual sugar mean? Why is fine wine aged in casks?

Teaching Suggestions

• Students are always interested in the process of making bread, beer, and wine (particularly the latter). I spend time talking about how to make bread, and what processes are happening during the proofing of the yeast, kneading, rising and baking. I'm amazed at the number of students who don't have the slightest idea how to make bread "from scratch." Since I made wine while in graduate school, and still brew my own beer, we go through the steps. Again, it's a topic that is sure to get their undivided attention.

Suggested Readings

Keister, E. "Prophets of gloom." *Discover*. November 1991. 53-56. Lichens, and the effects of atmospheric pollution.

Cherfas, J. "Disappearing mushrooms: another mass extinction." *Science*. 6 December 1991. 254: 1458. Europe's mushrooms are disappearing.

Janerette, C.A. "An introduction to mycorrhizae." *American Biology Teacher*. January 1991. 53 (1): 13-19.

Newhouse, J.R. "Chestnut blight." *Scientific American*. July 1990. 106-111.

Sokolov, R. "Subterranean treasures." *Natural History*. January 1991. 80-83. A description of truffles, along with a recipe.

Chapter 23. Plant Life

Chapter Overview

The various organisms classified in Kingdom Plantae are primarily photosynthetic autotrophs containing chlorophylls and carotenoids, and probably evolved from green algae, with which they share many characteristics. Plants are characterized by alternation of generation, in which the haploid gametophyte plant produces gametes, which after fertilization forms the diploid sporophyte plant. The sporophyte plant produces spores by meiosis, and these grow by mitosis into the gametophyte plant.

The bryophytes and allies are nonvascular, and are therefore small in size, and since their sperm are flagellated, must live in moist areas. In mosses, the dominant plant is the gametophyte, upon which the sporophyte lives. The ferns and allies are vascular plants, but do not produce seeds, although may produce differentiated spores (heterospory). The dominant generation is the sporophyte, which produces spores by meiosis, which grow into the smaller gametophyte. They also rely on water for sperm to swim through. Horsetails, whisk ferns and club mosses were once much more speciose, and are now represented by relatively small plants that live in moist habitats.

The evolution of seeds was a major step. Seeds develop from and are nourished by tissue of the female gametophyte, and contain multicellular embryos. Gymnosperms have naked seeds borne in cones, and angiosperms have seeds protected in fruits, which also aid in their dispersal. In both gymnosperms and angiosperms, the sporophyte generation dominates.

Gymnosperms include the conifers, ginkgos, gnetophytes and cycads. The last three groups were much more common in the Paleozoic and Mesozoic Eras. Conifers are the most speciose division, are typically evergreen and bear both male and female cones, typically on the same plant. The tree is the sporophyte, and the microspores and megaspores are produced in the cones. The microspores develop into the male gametophyte, which is the pollen grain. The megaspores develop into the female gametophyte, which remains within the cone. The pollen grain is transferred to the female cone, and grows a tube for the sperm to travel through. The sperm fertilizes the egg, forming an embryo which will grow into the sporophyte tree.

The angiosperms (Division Anthophyta) are the dominant plants now, including the mostly herbaceous monocots and the dicots, which may be woody. The seed of angiosperms is enclosed in a fruit, which develops from the ovary. Seeds also contain nutritive material, endosperm. We depend heavily upon flowering plants for our food (directly or indirectly), wood, fibers, drugs etc.

Lecture Outline

<u>Kingdom Plantae</u>

- Characteristics of plants
  - Eukaryotic
  - Size range from duckweed to sequoias
- Photosynthetic autotrophs
  - Contain chlorophyll *a* and *b*, and accessory pigments, the carotenoids
- Ancestors were probably the green algae
  - Share photosynthetic pigments
  - Both have cellulose-containing cell walls
  - Store carbohydrates as starch
- Adaptations to life on land
  - Waxy cuticle
  - Stomata allow gas exchange necessary for photosynthesis
- Sexual reproduction involves alternation of generation
  - Haploid portion of life is the gametophyte generation
    - Gives rise to gametes by mitosis
    - Has gametangia which produce gametes
      - Female gametangia produce eggs
      - Male gametangia produce sperm
      - Fertilization results in the zygote
      - Zygotes divide by mitosis and form the multicellular embryo
      - Embryonic development occurs within the female gametangia
      - Embryo develops into the sporophyte plant
  - Diploid portion of life is the sporophyte generation
    - Produces spores by meiosis
    - Spores develop into the gametophyte plant
- Four major groups of plants, three have vascular tissue
  - Vascular tissue consists of xylem and phloem
  - Xylem carries water and dissolved minerals
  - Phloem carries dissolved sugars

<u>Nonvascular plants</u>

- Bryophytes include the mosses, liverworts, and hornworts
- Lack vascular tissue
- Typically small in size
- Mosses live in dense colonies; individual plants are very small
  - No true roots; have rhizoids
  - Have stemlike and leaflike structures
  - Important in holding soil, in moist environments
  - Commercial importance: *Sphagnum*, the peat mosses
- Not in an evolutionary line with higher plants

- Alternation of generation
    - Have a dominant gametophyte generation (unusual)
    - Gametangium is at the top of plant; sexes are separate
        - Sperm are flagellated, swim to female gametangia during rains
    - Fertilization produces a diploid zygote
    - Zygotes develop by mitosis into sporophytes
        - The sporophyte grows out of the gametophyte
        - Derives nutrition from gametophyte
        - 3 parts: foot, seta (stalk), and capsule
    - Spores form by meiosis in the capsule
    - Spores disperse and grow into the gametophyte

<u>Seedless vascular plants</u>

- Evolved 400 million years ago
- Common in moist habitats, tropics
- Include ferns, whisk ferns, club mosses, horsetails
    - Typically small, but tree ferns in the tropics are very tall
- Have xylem and phloem
- Division Pterophyta, the ferns
    - Alternation of generations
    - Dominant plant is the sporophyte generation
        - Has an underground stem (rhizome), with roots and leaves (fronds)
    - Spores are produced in sporangia by meiosis
        - Sporangia are typically clustered in sori on the back of the fronds
    - Spores germinate and grow into the gametophytes by mitosis
    - Gametophytes are small, lack vascular tissues
    - Gametophytes have male and female gametangia
        - Gametangia produce gametes
        - Sperm swim through a film of water to the female gametangium
    - Diploid zygote grows by mitosis into a multicellular embryo
        - Embryo is dependent on the gametophyte at this stage
    - Embryo develops into the sporophyte; the gametophyte dies
- Fern allies
    - Whisk ferns (Division Psilophyta) are mostly extinct
    - Horsetails (Division Sphenophyta) were once dominant
        - Contributed to today's coal deposits
        - Grow in wet areas
        - Stems are high in silica, was used by pioneers to clean pots "Scouring rushes"
    - Club mosses (Division Lycophyta)
        - Mostly extinct
        - Also dominant earlier in the Paleozoic

Plants described so far produce only one type of spore
- Homospory

Some ferns and club mosses are heterosporous
- Microspores produce male gametophytes
- Megaspores produce female gametophytes

<u>The seed plants</u>

Seeds develop from the gametophyte and associated tissue

Compared to spores, seeds are advantageous
- Seeds contain a multicellular embryo
- Seeds contain a food supply
- Seeds have a protective seed coat

Two groups of seed plants
- Gymnosperms which produce seeds naked within cones
  - Example: pine trees
- Angiosperms (Anthophytes) produce seeds within a fruit
  - Example: the flowering plants

Both have vascular tissues

Also have alternation of generation, but gametophyte generation is much smaller and is dependent upon the sporophyte plant

<u>Gymnosperms</u>

Includes some of the world's tallest trees (redwood and sequoia) and oldest trees (bristlecone pines)

Four divisions which may share a common ancestor

Division Coniferophyta (conifers)
- Most speciose division
- Typically evergreen
- Leaves are needles
- Typically have male and female cones on the same tree
- Commercial uses:
  - Lumber
  - Paper
  - Turpentine and resins
- Life cycle
  - Tree is the sporophyte, produces microspores and megaspores in separate cones by meiotic division
  - Microspores develop into a tiny male gametophyte, the pollen grain
  - Megaspores are produced in the female cones, producing the female gametophyte, with eggs in the gametangia
  - Pollen arrives on the female cone, grows a tube towards the female gametangia
  - Sperm travel to the egg, forming a zygote
    - The zygote is surrounded by and is nourished by the female gametangia
  - The seed consists of the developing embryo and gametangia

Advantages of this mode of reproduction:
No need for water for the sperm; can live in dryer habitats
Sporophyte generation dominates, gametophyte is dependent on it

Conifer allies
Cycads (Division Cycadophyta) were once more common
Primitive tropical and subtropical plants
Very slow growing
Ginkgos (Division Ginkgophyta), the maidenhair tree
Extinct "in the wild," now only cultivated
Resistant to air pollution, often planted in cities
Gnetophytes (Division Gnetophyta) are an unusual group of advanced gymnosperms
Have vessels in their xylem, as do angiosperms

Division Anthophyta, the angiosperms

Most successful plants today
Seeds are enclosed in a ripe ovary, the fruit, which protects seeds
Alternation of generation, but gametophyte is very reduced
Double fertilization is unique; 2 nonflagellated sperm are involved
One sperm fertilizes the egg forming the zygote
The other sperm forms endosperm (nutritive tissue)
Includes our major food crops
Supply us with medicines, fibers etc.
Two classes: Class Monocotyledones and Class Dicotyledones
Monocots are typically herbaceous (nonwoody)
Have long narrow leaves with parallel veins
Floral parts in 3's, 6's
Have a single cotyledon (seed leaf), endosperm is persistent in the mature seed
Dicots may be woody
Have variable leaves, often broad, with branching veins
Have two cotyledons, and the endosperm is absorbed by the cotyledons
Angiosperms are extremely successful
Flowers facilitate pollination by attracting pollinators
Vessels in xylem and sieve tubes in phloem are very efficient in conduction
Leaves are efficient photosynthetic organs
Deciduous trees enter a dormant period during winter
Flowering plants are very adaptable, resulting in great diversity

Research and Discussion Topics

- Describe the past dominance by primitive plants like cycads and horsetails. Why do the gymnosperms and angiosperms dominate today?

• Describe our dependence on monocots in agriculture. List monocots that are typically used as human or livestock food.

Teaching Suggestions

• Overall, students tend to find plants a "dry" subject. I try to throw in lots of interesting facts, like the uses of different kinds of wood. White pines are a popular lumber tree. White spruce produces most newsprint, firs are used as Christmas trees. The berries of junipers are used to flavor gin. Various hardwoods (angiosperm wood) are used to make different things: find out what makes baseball bats, violins, bowling balls, furniture, balsa wood planes.

• Students also like to learn about plants that are edible. I discuss the following groups and their uses.

- Dicot Angiosperms
    - Nightshade Family
    - Carrot Family
    - Pumpkin Family
    - Sunflower Family
- Monocot Angiosperms
    - Grass Family
    - Lily Family
- Other Uses:
    - Seasonings
    - Dyes
    - Oils
    - Drugs and Narcotics:
        - Cocaine
        - Belladonna
        - Digitalis
        - Tobacco
        - Marijuana
        - Tea and Coffee

Suggested Readings

di Silvestro, R. "Lost in time." *International Wildlife*. July/August 1994. 42-45. A description of the oldest intact forest ecosystem in N. America in Ontario composed of ancient cedars; annual rings only composed of two cells.

Kendler, B.S., H.G. Koritz, A. Gibaldi. "Introducing students to ethnobotany." *American Biology Teacher*. January 1992. 54 (1): 46-50. Scientific investigation of plant use, including an assay using brine shrimp; with a great bibliography.

Lane, M. (and 8 others). "Forensic botany." *Bioscience*. January 1990. 40 (1): 34-39.

Delwiche, C.F., L.E. Graham, and N. Thomson. "Lignin like compounds and sporopollenin in Choleochaete, and algal model for land plant ancestry. *Science*. July 1989. 399-401. Another approach to the evolution of land plants.

Chapter 24. Animal Life: Invertebrates

Chapter Overview

The members of Kingdom Animalia include vertebrates and invertebrates. Invertebrates are much more speciose, as well as numerous. Animals evolved in marine habitats, and many have remained there, but members of many phyla are found in freshwaters and terrestrial habitats. Animals can be classified by their symmetry (or lack of symmetry) into the sponges, the radiate and bilateral animals. They can also be classified by the lack of a true body cavity (the sponges and flatworms), the possession of a pseudocoelom, or a true coelom (the coelomates). Further, they may be classified based on the first embryonic orifice. In protostomes, the first opening, the blastopore develops into the mouth, in deuterostomes, the first opening develops into the anus.

Sponges may be characterized as filter feeding, nonsymmetrical organisms with very simple bodies. The cnidarians have radial symmetry, and unique cnidocytes with nematocysts which allow them to be predators. Cnidarians are characterized by body forms of polyp or medusa. Examples of cnidarians include jellyfish, sea anemones and corals. Ctenophores (comb jellies) are similar, but lack cnidocytes. Members of Phylum Platyhelminthes are bilateral, cephalized organisms, and include free-living flatworms, parasitic flukes and tapeworms. These organisms are highly adapted as endoparasites and are extremely fecund.

Ribbon worms are of evolutionary significance, as they have developed a pseudocoelom, and a complete digestive tract. They also have a separate circulatory system. Nematodes are also pseudocoelomate worms, and are very speciose and numerous. Many are of agricultural importance, but the parasitic nematodes always attract attention. Rotifers are also pseudocoelomate, and are cell constant animals as well, of interest to developmental biologists.

Phylum Mollusca has the second greatest number of species (second only to arthropods). They range in size up to the largest of the invertebrates, the giant squid. The body plan of molluscs is coelomate and is characterized by a modified foot, mantle, visceral mass, and a unique feeding structure, the radula. Gastropods are the most speciose and diverse molluscs, and are the only ones to invade the terrestrial habitat. Bivalves exhibit reduced cephalization and are often filter feeders. Cephalopods have a closed circulatory system and are the most advanced molluscs.

The annelids are also coelomate, and exhibit segmentation in their body tube, their musculature and excretory systems. Polychaetes are marine worms with enlarged parapodia and many setae. They are often tube dwellers. Earthworms are deposit feeders, and lack parapodia and have few setae. Leeches are blood-sucking ectoparasites and lack both parapodia and setae.

Arthropods are the most successful group, and are characterized by a hard exoskeleton, paired jointed appendages, segmentation, specialized respiratory structures and well developed nervous systems. Arthropods may have evolved monophyletically from an annelid ancestor, or it may represent a polyphyletic group (insects and allies may have evolved from the Onychophorans). The chelicerates are characterized by chelicerae and pedipalps (most are predators), and include spiders, scorpions, ticks, mites. Crustaceans, sometimes called the insects of the sea, are mandibulate and are extremely numerous in aquatic habitats. They differ from insects in the possession of biramous appendages and two pairs of antennae.

Insects are the terrestrial mandibulates, have uniramous appendages and one pair of antennae. Predatory centipedes and herbivorous millipedes are closely related to insects. Insects are characterized by their 6 legs, tracheal respiratory system, wings in most, and small size. Insects have separate sexes, internal fertilization, and may be characterized by gradual or complete metamorphosis.

Echinoderms are deuterostome invertebrates found in marine habitats. They have a pentaradial symmetry, and an epidermis covering their endoskeleton. They have rather rudimentary systems, but have a unique water vascular system including tube feet.

Lecture Outline

<u>Characteristics of animals</u>

- Multicellular heterotrophs; the majority of which are invertebrates
- Classified in approximately 35 phyla
- Except in simplest animals, cells are specialized into tissues and organs
- Food is typically ingested, then digested in a digestive cavity
- Most are motile, at least in some stage of life
- Typically, animals have well-developed sense organs and nervous systems
- Most reproduce sexually
- Animals are ubiquitous; they are found nearly everywhere
    - The marine habitat is most hospitable
        - Salt water is nearly isotonic with most invertebrates
        - Water buoys animals
        - Temperature is relatively constant
        - Lots of animals; therefore lots of food for predators
    - Freshwater habitats are hypotonic to body fluids
    - Terrestrial habitats pose the problem of dehydration

<u>Classifications of animals</u>

- Classification based on body plans
    - Sponges have no symmetry, may be classified in subkingdom Parazoa
    - Other animals have symmetry, classified in subkingdom Eumetazoa
        - Organisms which are primarily radially symmetrical are the cnidarians and ctenophores

- Not cephalized
- All other animals are bilaterally symmetrical
  - Sea stars and relatives look radial; evolved from bilateral ancestors
  - Bilateral symmetry is an advantage for motility
  - Increasing cephalization seen in more complex animals
- Bilaterally symmetrical animals' parts may be described as dorsal, ventral, lateral, medial, posterior and anterior
- Bilateral animals may be divided into 3 planes: sagittal, frontal, and transverse

- Animals may be classified based on body cavity structure
  - Acoelomate animals have no true coelom
    - Called the solid worms; have no body cavity other than the gut
  - Pseudocoelomate animals have a false coelom
    - A tube-within-a-tube arrangement
    - The pseudocoelom surrounds the gut
      - Formed between endoderm and mesoderm
  - Coelomate animals have a true coelom
    - The coelom forms entirely within mesoderm
- Animals may be classified based on embryonic development
  - In protostomes, the first opening in the blastula is the blastopore
    - The blastopore develops into the mouth
    - Includes all invertebrates except echinoderms
  - In deuterostomes, the first opening in the blastula develops into the anus, another opening formed later becomes the mouth
    - Includes the chordates and the echinoderms

<u>Phylum Porifera- the sponges</u>

- Most primitive animals, mostly marine, typically sessile
- Structured by spicules and the protein spongin
- Feed by filter feeding; water passes into the central spongocoel
  - Water flows out the apical osculum
- No tissues, but cells do divide tasks; collar cells trap food, ameba-like cells distribute food
  - Digestion is intracellular; there is no gastrointestinal cavity
- Exchange of gases is by diffusion
- Reproduction may be sexual (larvae are ciliated and motile) or asexual
  - Most sponges are hermaphroditic; one sponge produces eggs and sperm

<u>Cnidarians</u>

- Most are marine
- 3 classes: the hydras and Portuguese man-o-war (Class Hydrozoa), the true jellyfish (Class Scyphozoa), and the sea anemones and coral (Class Anthozoa)
- Cnidocytes are unique cells with nematocysts which may discharge to ensnare and sting prey

- Digestive system has one opening (incomplete digestive tract)
  - Digestive cavity is the gastrovascular cavity
- Two tissue layers; outer epidermis, and inner gastrodermis
  - Mesoglea in between contains a few cells
- Gas exchange is accomplished by diffusion
- Have a nerve net; impulses pass in all directions equally
- Two body forms: polyp and medusa
  - Polyp resembles hydra, medusa resembles a jellyfish
  - See both body forms in many cnidarians
- See alternation between sexual and asexual stages
  - Different from alternation of generation; all forms are diploid except gametes
  - Larvae is a planula
- Hydra
  - Tiny predator, often reproduce asexually by budding
- Jellyfish
  - Medusa is the adult body form
- Corals and sea anemones
  - No medusa stage
  - Coral reefs are formed by coral colonies and coralline algae
    - Rely on symbiotic algae (green algae and dinoflagellates)

Phylum Ctenophora, the comb jellies

- Previously lumped with the cnidarians
- Rows of cilia that look like combs
- Similar body form to a medusae
- Lack cnidocytes but have tentacles with adhesive

Phylum Platyhelminthes, the flatworms

- Includes free-living flatworms (the turbellarians), the flukes, and the tapeworms
- Have three tissue layers; ectoderm, mesoderm and endoderm
- Have organs
- Are acoelomate, bilaterally symmetrical, cephalized
- Nervous system includes ganglia and a ladder-like set of nerves
- Freshwater flatworms have protonephridia with flame cells to excrete water
- Incomplete digestive cavity (gastrovascular cavity)
  - Mouth leads to pharynx then branched gastrovascular cavity
  - No digestive cavity in tapeworms
- Free-living flatworms
  - Examples: *Dugesia*, some other genera found in moist gardens, others marine
  - Reproduce asexually or sexually
    - Can regenerate lost parts, reproduce also by dividing in half
    - Sexual planarians are hermaphroditic

Flukes are parasites, in the blood, liver or lungs of host
- Alternate sexual and asexual stages
- Often have one stage of life in fish or snails
- Example: *Schistosoma*, which causes schistosomiasis in tropical areas

Tapeworms are the ultimate parasite
- Long worms, are endoparasites in the intestines of vertebrates
- No head; have a scolex with suckers and hooks
- Segments (proglottids) have both male and female reproductive organs
  - Reproductive capacity is amazing
- Lack digestive organs; they <u>live</u> in one!
- Humans become infected when eating poorly cooked beef, pork or fish

<u>Phylum Nemertea, the ribbon worms</u>

- Mostly marine, long, predatory worms
- Also called the proboscis worms; use an eversible proboscis to capture prey
- Have a complete digestive tract; mouth to anus
  - Tube-within-a-tube; a pseudocoelomate
- Have a separate circulatory system, but no heart

<u>Phylum Nematoda, the nematodes or roundworms</u>

- Extremely speciose, numerous, and widely distributed
- Many are free-living, others are parasites in animals and plants
- Are pseudocoelomate, have a complete digestive tract, but lack a circulatory system
- Example: *Ascaris*, the intestinal roundworm
  - Lives in the intestines of humans
  - Enormous reproductive output; sexes are separate

<u>Phylum Rotifera, the rotifers or wheel animals</u>

- Also pseudocoelomate; very small, mostly freshwater
- Have a corolla of cilia; brings food into the pharynx where food is ground up
- Are cell constant, as are nematodes
  - Cell division during embryonic division is highly regulated; mitosis does not take place in the adult
  - Growth and repair are not possible; neither is cancerous growth

<u>Phylum Mollusca</u>

- Second largest phylum, with respect to species numbers
- Includes the largest invertebrates; the giant squid and the giant clam
- Mostly marine, but snails and clams may be freshwater, snails are also terrestrial
- Body plan:
  - Foot, for locomotion
  - Visceral mass houses the organs
  - Mantle cloaks the visceral mass, may form the shell, if present
  - Most have a radula, a rasplike structure for grazing
  - Open circulatory system except in the cephalopods

Class Polyplacophora; the chitons have 8 plates
- Herbivores, common in the intertidal

Class Gastropoda; the snails and slugs
- Marine, freshwater or terrestrial
- Highly cephalized, often with a single coiled shell
  - Some lack shells (slugs and nudibranchs)
  - Limpets have noncoiled shells
- Due to coiling of shell the embryo rotates, called torsion
  - Unfortunately, the anus tends to dump on top of the head and gills

Class Bivalvia; the clams, mussels and oysters
- Have a two part shell
- May form pearls
- Foot is muscular for digging in many species, strong muscles hold the two shells together
- Sexes are usually separate; fertilization typically external, but within the gill cavity of the female in some
- Larvae is a trochophore larvae, developing further into a veliger
  - Both are readily eaten by small marine predators and filter feeders

Class Cephalopoda; the squids, cuttlefish, octopods, nautiluses
- Active predators, tentacles to capture prey
- Ancestors had shells, nautilus has a shell, squid have reduced shells, octopods have no shells

<u>Phylum Annelida, the segmented worms</u>

Coelomate worms with segmented bodies
- Segments are separated by septa; muscles and excretory organs (nephridia) are segmented as well
- Segmentation facilitates locomotion; each segment has own muscles

Setae aid in locomotion

Closed circulatory system and a complete digestive tract

Respiration is via gills or cutaneous

Nervous system consists of an anterior ganglia (brain) and ventral nerve cords
- Ganglia are seen in each segment

Class Polychaeta includes the sand worms and tube worms
- May be free-swimming or sedentary in tubes
- Have parapodia (fleshy appendages) with many setae

Class Oligochaeta, the earthworms and relatives
- Terrestrial or freshwater
- Lack parapodia, few setae
- Mostly hermaphroditic
- Includes *Lumbricus terrestris*, the common earthworm
  - Is a sediment feeder; eating dirt and organic matter

- Class Hirudinea, the leeches
    - Blood-sucking ectoparasites
    - Mostly freshwater; a few marine and terrestrial
    - Lack parapodia, setae
    - Used by humans in the past to remove blood, called leaching
        - Now being used to remove pooled blood in grafts, etc.
        - Inject hirudin (an anticoagulant), an anesthetic, and antibacterial agent into the incision

<u>Phylum Arthropoda</u>

- The most speciose and numerous animal phylum
- Success due to:
    - Paired, jointed appendages, specialized for feeding, walking, swimming, and reproduction
    - Exoskeleton, made of chitin and protein, prevents water loss, aids in protection against predators, anchoring sites for muscles
        - Disadvantageous because it does not grow with the animal
    - Segmented body
        - Fusion of segments for specialized functions
        - Head, thorax and abdomen
    - Specialized respiratory systems; gills, trachea
    - Well-developed nervous systems, complex behaviors
- Have an open circulatory system
- Share a common ancestor with the annelids
    - Share segmentation, nervous system design
- All arthropods may have a common ancestor, or may have had a polyphyletic origin (crustaceans and chelicerates had a common ancestor, uniramians had a separate ancestor)
    - Uniramids may have evolved from onychophorans
    - Are both internally segmented, but jaws are different
    - Both groups have trachea for respiratory tubules
- Here, divide into 3 subphyla: chelicerates, crustaceans, and insects and allies

<u>Chelicerate arthropods</u>

- Subphylum Chelicerata
- Horseshoe crabs, spiders, scorpions, ticks and mites
- Lack antennae and mandibles
- Chelicerae are unique mouthparts, along with pedipalps (food handlers)
- Class Merostomata
    - Horseshoe crabs and extinct eurypterids
- Class Arachnida
    - Typically predatory
    - Six pairs of jointed appendages
    - Spiders with silk glands and spinnerets
        - Poison glands, few are dangerous to humans
            - Black widow spider has a neurotoxin
            - Brown recluse also venomous

Mites and ticks are economically harmful
May cause bites, plant infestations
Carry Lyme disease and Rocky Mt. spotted fever

Crustacean arthropods

Subphylum Crustacea includes lobsters, crabs, shrimp and barnacles
Often are primary consumers in aquatic food chains
Biramous appendages, two pair of antennae, mandibles as mouthparts
Typically have 5 pairs of walking legs
Barnacles are unusual crustaceans- Linnaeus thought they were bivalves!

Insects and their allies

Subphylum Uniramia includes insects, centipedes and millipedes
One pair of antennae, uniramous appendages
Centipedes have one pair of legs per segment, millipedes have two
Centipedes are carnivorous, millipedes are herbivorous or scavengers
Class Insecta
Evolved about 360 million years ago in the Devonian Period
Highly successful, both in terms of numbers of individuals and species
+750,000 species currently identified
Primarily terrestrial
Have trachea for respiration
Openings to the outside- spiracles
Allows very high metabolic rate
Head with highly adapted mouthparts
Excretory system consists of Malpighian tubules
Reproduction
Sexes are separate, internal fertilization
During development, molts allow growth
Incomplete metamorphosis (gradual development) with several nymphal stages
Complete metamorphosis with egg, larva, pupa and adult stages
What allows the success of the insects?
Adaptability of body plan
Flight
Small size
Metamorphosis prevents competition between larvae and adults

Phylum Echinodermata, the echinoderms

Are deuterostomes with pentaradial symmetry as adults
Larvae are bilateral; they evolved from bilateral ancestors
Have a thin epidermis covering their endoskeleton
Unique water vascular system
Tube feet for locomotion, assisting in feeding, gas exchange
Well developed coelom, complete digestive system
Various respiratory structures, circulatory system is not well developed
No excretory structures; nerve rings are radial, simple

Research and Discussion Topics

• Discuss the current use of *Caenorhabditis elegans,* a tiny roundworm in developmental and genetic research. What are the advantages to using this roundworm, compared, for example, to the use of fruit flies?

Suggested Readings

Turpin, F.T. "The insect appreciation digest." 1992. Published by the Entomological Foundation (9301 Annapolis Rd., Suite 300, Lanham MC 20706.) This inexpensive book describes the study of insects, and their links to humans. Very humorous.

Vollrath, F. "Spider webs and silks." *Scientific American*. March 1992. 70-76. Silk production and web designs.

Scanlon, K. "Behold the mite." *Equinox.* March/April 1992. 34-39. Reprinted in SIRS Science, 1992, article 64. This must be the funniest article I have ever read; a description of the mites that live on and with us. Included are the effects of mite fecal material and asthma, and mites in our beds which live on our shed skin cells and dried semen.

Anonymous. "Entomologists wane as insects wax." *Science.* November 1989. 754-756. Jobs in entomology.

Boyle, R.H. "The joy of cooking insects." *Audubon.* Sept-Oct. 1992. Spurred by the New York Entomological Society's 100th anniversary insect "dinner," a description of insects that people have historically eaten, and do presently eat.

Various authors. Special issue of *Natural History* devoted to mosquitoes. July 1991.

Conniff, R. "The little suckers have made a comeback." *Discover.* August 1987. 85-94. A description of leeches, their past and present use in medicine, ecology.

Bunkley-Williams, L. and E.H. Williams. "Global assault on coral reefs." *Natural History.* April 1990. 49-54. A look at coral bleaching; symptoms and causes.

Roberts, L. "Coral bleaching threatens Atlantic reefs." *Science.* 27 November 1987. 238: 1228-1229.

Chapter 25. Animal Life: Chordates

Chapter Overview

Phylum Chordata includes three subphyla; two which are invertebrates and Subphylum Vertebrata which include the familiar vertebrates. All share the basic chordate characteristics at some point in their life: notochord, dorsal hollow nerve cord, pharyngeal gill clefts and postanal tail. In the urochordates, the tunicates, the adults only retain the gill slits; in the cephalochordates, the lancelets have all four characteristics.

The vertebrates are characterized by an endoskeleton; the vertebrae which enclose the nerve cord, and the cranium which encloses the brain. Increasing cephalization, paired appendages, and advanced circulatory and digestive systems allowed the evolutionary success of this group. The jawless fish first appeared in the Ordovician period; the ostracoderms. The only extant jawless fish are the lampreys and hagfish. The jawed freshwater placoderms gave rise to the chondrichthyean and osteichthyean fish. The chondrichthyean fish are nearly all marine, and have cartilaginous skeletons, and include the sharks, skates and rays. The osteichthyean fish evolved in fresh water, and two major groups are extant; the lobe-finned fish, from which amphibians evolved, and the diverse and numerous ray-finned fish.

The amphibians became the first terrestrial vertebrates, but most return to water to breed. Their skin is smooth and scaleless, and many rely on cutaneous respiration. The reptiles, which evolved from amphibians, have scaly skin which precludes respiration, have internal fertilization, and lay shelled eggs on land. Both birds and mammals evolved from reptilian ancestors.

Birds are highly adapted for flight, with wings, feathers, an extremely light skeleton, a 4 chambered heart, and a very efficient respiratory system. Both birds and mammals are endothermic. Mammals evolved from therapsids, and the first mammals coexisted with the dinosaurs by being arboreal and nocturnal. After the Mesozoic Era, the mammals exhibited adaptive radiation, developing specialized teeth and locomotion. All mammals have hair at some point in their life, and have mammary glands. The monotremes lay eggs, but nurse the young after they hatch. The marsupials give birth to their young at a very undeveloped state, and the young nurse while in the marsupium. The placental mammals have a placental attachment between the female and the embryo, allowing the young to be born at a relatively well developed stage.

The primates evolved from tree shrews and possessed many adaptations because of their arboreal life; prehensile tails, opposable thumbs, 3-D vision, freely movable arms and legs. The primates may be divided into the prosimians and the anthropoids, which can be further divided into the New World and Old World monkeys. The apes and hominids evolved from the Old World monkeys.

The hominid line separated approximately 6 million years ago; *Australopithecus* was the first hominid genus, and lived in Africa. These ancestors had a bipedal stance, and an ever increasing brain size. *Homo habilis* was noted for consistent tool use. *H. erectus* evolved in Africa and moved into Europe and Asia and had complex tools, used fire and lived in caves or primitive shelters. From *H. erectus* evolved our species, *H. sapiens*. The Neanderthals lived in Europe and were primarily hunters, and were ultimately replaced by the Cro-Magnon culture. During the last 50,000 years, our species has passed from the hunter/food gatherer stage to the development of agriculture (10,000 years ago) to the Industrial Revolution.

Lecture Outline

<u>Phylum Chordata</u>

- Three subphyla; 2 are invertebrates
    - Subphylum Urochordata, the sea squirts
    - Subphylum Cephalochordata, the lancelets
    - Subphylum Vertebrata, the vertebrates
- Ancestor largely unknown
- Share four characteristics
    - Have a notochord at some point in life cycle
        - A flexible dorsal supporting rod
    - Dorsal hollow nerve cord
        - Nonchordate invertebrates have ventral nerve cords which are solid
    - Pharyngeal gill grooves or slits
        - Important in vertebrates as the gill supports formed the jaw
    - Postanal tail
        - Nonchordate invertebrates have a terminal anus
- Subphylum Urochordates, the tunicates or sea squirts
    - Marine, typically round in shape with a tunic containing the organs
    - Larvae have chordate characteristics, adults have only the gill slits
    - Filter feeders, may be colonial
- Subphylum Cephalochordata
    - Adults have all four chordate characteristics
    - Example: *Branchiostoma*
    - Lack paired fins, jaws, heart; therefore are not fish, although they resemble them
- Subphylum Vertebrata
    - Have a vertebral column which develops around the notochord and mostly replaces it (notochord present as intervertebral discs)
    - Braincase protects the brain
    - Have a living endoskeleton
    - Typically have two pairs of appendages
    - Greatly cephalized
    - Closed circulatory system with a 2, 3 or 4 chambered heart

Complex digestive system
Sexes are almost always separate

Class Agnatha

The jawless fish; the lampreys and hagfish
Evolved from armored jawless freshwater fish in the Devonian
Lampreys may be temporal ectoparasites on other fish

Class Chondrichthyes

Evolved from armored jawed fish with paired fins, the placoderms
Includes sharks, skates and rays
Have cartilaginous skeletons
Have placoid scales which resemble their teeth in structure
Many are predators, but the largest sharks are filter feeders
Skates and rays are flattened and typically are benthic

Class Osteichthyes

The bony fish; have osseous skeletons
The most speciose fish
Have flexible scales
Have paired and unpaired fins
Operculum covers the gills
Most are oviparous
Ray-finned fish are most complex and speciose
Lungs were modified as swim bladders
Lobe-finned fish gave rise to the amphibians
Lungs were adaptive as they made the switch to terrestrial life

Class Amphibia

Includes frogs, toads, salamanders, newts, caecilians
Are quasi-terrestrial, typically returning to water to lay eggs
Cutaneous respiration is important
Three-chambered heart

Class Reptilia

Evolved from a labrinthodont amphibian
Examples: turtles, snakes, lizards and crocodilians
Included the dinosaurs, which showed adaptive radiation during the Mesozoic Era
Birds and mammals evolved from reptilian ancestors
Internal fertilization, shelled eggs allow a terrestrial life
Egg contains yolk, amniotic fluid, other membranes
Scaly skin prevents desiccation, must be shed
Are ectothermic
Lungs are more developed than amphibian lungs, heart is three chambered or four chambered (crocodiles and alligators)

Class Aves

- Birds evolved from lizard-like dinosaurs
- *Archaeopteryx* shows intermediate characteristics; teeth, claws on wings
- Birds have feathers, scales on legs, lay eggs
- Show many adaptations for flight, wings, feathers, light bones
- Flightless birds include penguins, ostriches, cassowaries
- Efficient respiratory system including air sacs which extend into the bones
  - Four chambered heart
- Endothermic

Class Mammalia

- Mammals evolved from therapsids in the Triassic period
- First mammals coexisted with the ruling reptiles by being small, nocturnal, and arboreal
- Adaptive radiation during the Cenozoic Era; specializations in teeth, locomotion
- All have hair at some point in life cycle
- All have mammary glands
- Monotremes lay eggs
  - Examples: duck-billed platypus and echidna (spiny anteater)
  - Nurse young after they hatch
- Marsupials have a pouch (marsupium)
  - Examples: kangaroo and koala
  - Early embryonic development occurs in the uterus; nourished by yolk
  - After a few weeks, the young are born and crawl to the pouch, where they attach to a nipple
  - Most common in Australia; only the opossum is common in the Americas
  - Marsupials show parallel evolution with the placentals elsewhere
- Placental mammals are the most common mammals
  - Embryos develop in the uterus, placenta is both maternal and embryonic tissue, and is the site of exchange of gases and nutrients
  - 17 orders of extant mammals

Order Primates

- Humans, monkeys, apes, lemurs belong to this group
- Evolved from tree shrews during the Mesozoic
- Important adaptations because ancestors were arboreal
  - Long, freely movable limbs
  - Opposable thumb
  - Eyes at the front of the head allow 3-D vision
  - Acute vision
- Highly developed social behaviors
- Two suborders
  - Prosimians- the lemurs, lorises and tarsiers

- Anthropoids- the monkeys, apes and humans
  - Evolved from prosimians in the Oligocene Epoch
  - New World monkeys are found in S. America
    - Smaller thumb, may be absent
    - Tails may be prehensile
    - Live in social groups
    - Includes howler, squirrel and spider monkeys
  - Old World monkeys are found in Africa and Asia
    - May be arboreal or ground dwellers
    - Ground dwellers are quadrupedal
    - Tail is never prehensile, some lack tails
    - Opposable thumbs
    - Nostrils are directed downward
    - Are larger monkeys, are social

<u>Hominoids</u>

- Hominoids includes apes and hominids (humans and their ancestors)
  - Gibbons are in one ape family
    - Arboreal, brachiate for locomotion
  - Family Pongidae includes orangutans, gorillas and chimpanzees
    - Orangutans are arboreal
    - Gorillas and chimps are terrestrial
    - Complex social behavior
    - Molecular evidence shows close relationship to humans
  - Hominids evolved approximately 6 million years ago
    - Developed bipedal posture, enlarged brain
    - Evolved originally in Africa; first genus was *Australopithecus*
    - *A. afarensis* --> *A. africanus*
    - First *Homo* was *H. habilis* over 2 million years ago
      - Consistently used tools
    - *H. erectus* migrated into Europe and Asia
      - Was fully upright, taller, advanced tools, clothing
      - Were probably scavengers, no weapons unearthed
    - *H. sapiens* appeared 200,000 years ago
      - Neandertals were hunters, had religious beliefs of some type, don't know why they disappeared
    - First *H. sapiens sapiens* was the Cro-Magnon culture in Europe
      - Cave art, complex weapons and tools
      - Most likely had language; important in generational transmission of culture
    - Cultural evolution
      - Development of hunter/food gatherer societies
        - Division of labor
        - Hunting declined approximately 15,000 years ago

Development of agriculture
- Approximately 10,000 years ago
- Developed independently in Asia and Americas
- Crop cultivation, then animal domestication
- Aggregations of people formed villages
- Allowed development of craftsmen

Industrial revolution

Research and Discussion Topics

• Assess the various theories concerning the extinction of the dinosaurs and other Mesozoic reptiles. Which has the most geological and biological support?

• Discuss the differences and similarities in the development of agriculture in the Middle East and the Americas. Grains like wheat were the staple of the people of the Middle East, whereas corn was the staple in the Americas. How did the differences in growing and harvesting these crops result in differences in the cultures?

Suggested Readings

Wilson, A.C. and R.L. Cann. "The recent African genesis of humans." *Scientific American*. April 1992. 68-73. An investigation of the maternal ancestor and mitochondrial DNA.

Thorne, A.G. and M.H. Wolpoff. "The multiregional evolution of humans." *Scientific American*. April 1992. 76-83.

Horgan, J. "Early arrivals." *Scientific American*. February 1992. 17-20. A description of the controversy surrounding the date our species arrived in the New World.

Shreeve, J. "Argument over a woman." *Discover*. August 1990. The search for "Eve" using mitochondrial DNA.

Stringer, C.B. "Fate of the Neanderthal." *Natural History*. December 1984. Speculations on the disappearance of the Neanderthals.

Fricke, H. "Coelocanths: the fish that time forgot." *National Geographic*. June 1988. 824-838.

Alexander, R.M. "How dinosaurs ran." *Scientific American*. April 1991. 130-136. Physical engineering gives clues to how fast the large dinosaurs may have run.

Alvarez, W. and F. Asaro. "An extraterrestrial impact." *Scientific American*. October 1990. 78-84. Evidence for an asteroid which caused the Cretaceous extinctions.

Robbins, J. "The real Jurassic park." *Scientific American*. October 1990. 52-59. Bakker's theories about the extinctions.

Konishi, M., Emlen, S.T., R.E. Ricklefs, and J.C. Wingfield. "Contributions of bird studies to biology." *Science*. 27 October 1989. 465-472. A discussion of bird studies in behavioral, developmental biology, physiological and evolutionary studies.

Chapter 26. Plant Structure

Chapter Overview

Plants, particularly flowering plants, are extremely diverse in form, but share a similar body form. Herbaceous plants lack secondary growth and often live and reproduce in a year or two. Woody plants are perennials which increase in width by addition of secondary growth.

The plant body may be divided into the root and the shoot system. Both contain differentiated tissues and the vascular tissue is continuous between the two systems. Both contain meristematic tissue. Root systems anchor the plant and conduct water and dissolved minerals up to the above-ground portion of the plants. Stems support the leaves, the primary photosynthetic organs of the plant.

Plant tissues make up the organs of the plant. The ground tissue system is composed of parenchyma, which function in photosynthesis and storage, and collenchyma and sclerenchyma which have thickened cell walls and function in strengthening herbaceous plants. Vascular tissue, composed of xylem and phloem, is a complex tissue which conducts fluids. Their tube-like cells are continuous throughout the plant. Dermal tissues cover the plant. In herbaceous plants, the epidermis is very thin. In woody plants, cork and cork parenchyma cells compose the bark.

Plants exhibit growth at meristems. Both roots and stems have apical meristems, composed of cells which undergo mitosis and differentiation. Root meristems are protected by the root cap. Stem meristems are protected by the leaf and bud primordia. Apical meristems contribute primary growth to a plant. Secondary growth takes place at lateral meristems, composed of the vascular cambium and the cork cambium.

Lecture Outline

<u>Plant diversity and similarity</u>

- 235,000 species of flowering plants
- Herbaceous plants
  - No woody tissue
  - May be annuals (live and reproduce in one year)
  - May be biennials (live and reproduce in a two year span)
  - May be perennials (live many years, often die back each year)
- Woody plants
  - All are perennials
  - May shed leaves annually and become dormant over winter
- Plant bodies are organized into root and shoot systems
  - Roots are below ground, exposed to dark, most soil

Shoot system is typically above ground, exposed to dry air, sunlight
- Consists of stem with leaves, flowers, fruits

Root systems

- Underground portion of plant
- Functions to anchor plant, absorb nutrients and water
- Taproot system
  - One main root with lateral roots coming from it
  - Gymnosperms and dicot plants
  - Develops from embryonic root
  - Example: dandelion
- Fibrous systems
  - Has many roots of similar size, lateral branches from them
  - Monocots have fibrous roots
  - Example: the little roots seen on green onions

Shoot system

- Stem with leaves
  - Typically just the leaves are photosynthetic
  - Leaves are attached at nodes
  - An internode is the stem between nodes
- Buds are embryonic shoots
  - Terminal buds are at the end of the stem
  - Lateral buds are located in leaf axils
    - Buds are covered by bud scales (modified leaves)
    - When buds resume growth, bud scales fall off, leaving bud scale scars
  - Can age a branch by counting the bud scale scars of temperate zone plants
  - Lenticels are also seen on woody stems
    - Openings to allow diffusion of gas into stem
- Leaves
  - Blade and petiole (stalk)
  - Simple leaves have no divisions; compound leaves are divided into leaflets
  - Lateral buds form at the base of the leaf; can use the presence of lateral buds to distinguish between leaves and leaflets
  - Leaf arrangement:
    - Alternate- one leaf per node
    - Opposite- two leaves per node
    - Whorled- more than two leaves per node
  - Leaf veins:
    - Parallel veins are characteristic of monocots
    - Netted veins are characteristic of dicots
      - May be palmately netted (radiate from one point)
      - May be pinnately netted (branch off entire length of major vein)

Plant tissues

- Cells are organized into tissues
- Simple tissues have only one cell type
- Complex tissues have two or more types of cells
- Three tissue systems in plants:
    - Ground tissue
    - Vascular tissue
    - Dermal tissue
- Cell walls
    - Exterior to plasma membranes
    - Primary cell wall is thin, produced first
    - Secondary cell wall is made after the cell stops growing
        - Located between primary wall and plasma membrane
- Ground tissue
    - Composed of:
        - Parenchyma tissue
            - Made of parenchyma cells only (a simple tissue)
            - Function in photosynthesis, storage and secretion
            - Cells are living
            - Make up most of the plant body
        - Collenchyma tissue
            - Made of collenchyma cells (a simple tissue)
            - Function in structural support, particularly along leaf veins and the surface of the stem
                - Especially important in herbaceous plants
            - Cells are alive
            - Primary cell walls are thickened in the corners
        - Sclerenchyma tissue
            - Functions in support
            - Cells are dead at maturity
            - Thickened primary and secondary cell walls
            - Fibers are abundant in wood and bark, nut shells
            - Also impart the gritty texture to pear fruits
- Vascular tissue consists of xylem and phloem
    - Is a complex tissue
    - Function: conduction of materials, provides structural support
    - Carry fluid upwards in the plant, from roots to the shoot system
    - Xylem is composed of:
        - Tracheids
            - Cells are dead at maturity
            - Conduct water and dissolved minerals
        - Vessel elements
            - Only seen in flowering plants
            - Cells are also dead at maturity
            - Also conduct water and dissolved minerals
            - Typically wider than tracheids
            - Have holes at the ends; are stacked into vessels

- Both tracheids and vessel elements have pits to allow lateral transport of fluids
- Parenchyma cells
  - Cells store materials
- Fibers
  - Provide structural support

Phloem is composed of:

- Sieve-tube members
  - Extremely specialized cells for conduction of sugar solution
  - Have perforated ends called sieve plates
  - Are living at maturity, but many cellular organelles degenerate; can function without nuclei!
  - Typically live less than a year- palm trees are an exception
- Companion cells
  - Aids the sieve-tube member
  - Are alive and have functioning organelles
- Fibers
  - Many fibers provide support for the entire plant body
- Parenchyma cells

<u>Dermal tissue</u>

- Covers a plant
  - In herbaceous plants, may be only 1 cell thick (epidermis)
  - In woody plants, the epidermis splits and periderm makes the bark
- Epidermis
  - Consists of parenchyma, guard cells, and trichomes
  - Thick cell walls on the outer surface
  - Function in preventing water loss
    - Typically are not photosynthetic
  - Cuticle is the wax produced by the epidermal cells
  - Stomata are openings into the leaf
    - Guard cells form the stomata
    - Gases pass in and out by diffusion
    - Stomata are typically open in the day, closed at night
  - Trichomes
    - Root hairs are trichomes which increase the surface area of the roots
    - Other trichomes are specialized to secrete salt or noxious compounds
- Periderm forms in woody plants
  - Seen both on the root and shoot systems
  - Complex tissue made of:
    - Cork cells which are dead at maturity
      - Thick cell walls
    - Cork parenchyma cells function in storage

Growth in plants occurs at meristems

- Growth involves:
  - Cell division (mitosis); an increase in the number of cells
  - Cell elongation
  - Differentiation; specialization of cells
- Animals grow all over, plants just grow in meristems
  - Meristematic cells are not differentiated, and can divide by mitosis
  - Meristems retain this ability for the entire life of the plant
- Primary growth results in an increase in height (length) of the plant
- Secondary growth results in the girth of a plant
  - Seen only in gymnosperms and woody dicots
- Apical meristems
  - Found at the tips of roots and stems
  - Form primary growth
  - Root meristems:
    - Root cap at very end of root
    - Next is meristematic area; cells are small, rapidly dividing
    - Further up is the area of cell elongation
    - The most mature cells which have begun to differentiate follow
      - Root hairs may be seen in this area
  - Stem meristems:
    - Stem apical meristem is at center of tip of stem
    - Leaf primordia (embryonic leaves)
      - Cover the meristem
    - Bud primordia (embryonic buds)
- Lateral meristems
  - Form secondary growth
    - Vascular cambium is a lateral meristem
      - Tube of cells between the wood and bark
      - Cells divide, add more cells to wood and inner bark
    - Cork cambium
      - Irregular arrangement of cells in the outer bark
      - Cells divide, add more cork and cork parenchyma cells
      - Ultimately adding to the periderm

Research and Discussion Topics

• Discuss the differences between monocot palm "trees" and dicot trees like oak and maple trees. Describe the differences between their root systems and shapes of trunks. Which do you think would be easier to transplant, a 25 foot palm tree or a 25 foot oak tree?

Teaching Suggestions

• Relate the root meristematic area to the onion root tips which students may have seen in a laboratory.

Suggested Readings

Booth, W. "Combing the earth for cures to cancer, AIDS." *Science*. 28 August 1987. 969-970. Candidates for pharmaceuticals in the tropics.

Dussourd, D.E. "The vein drain; or, how insects outsmart plants." *Natural History*. February 1990. 44-49. Herbivorous insects and the plants they eat.

Chapter 27. Leaves

Chapter Overview

Leaves are the primary photosynthetic organ of the plant. The epidermis forms the boundary of the leaf, and the waxy cuticle helps to prevent water loss. Typically more numerous in the lower epidermis, stomata allow carbon dioxide to diffuse in, and oxygen to diffuse out. The photosynthetic cells of the leaf are between the epidermal layers, the mesophyll cells. The palisade layer(s) are under the upper epidermis and are the primary site of photosynthesis. The cells of the spongy layer are arranged loosely, to allow diffusion of gases. Veins are vascular bundles which branch throughout the mesophyll; xylem bringing water and dissolved minerals to the leaf cells, phloem carrying excess dissolved sugars to other parts of the plant.

Monocot leaves tend to have parallel veins, dumb-bell shaped stomata and lack differentiated mesophyll. Typical dicot leaves have netted venation, bean shaped stomata and mesophyll differentiated into palisade and spongy layers.

Stomata allow diffusion of gases, but significant loss of water as well. To prevent excess loss, stomata close at night, and during water stress. Stomates open in response to low carbon dioxide layers, and also automatically in response to light, based on a potassium ion/water influx. Although it represents a net loss, transpiration is also important as it cools the leaves, and concentrates nutrients within the plant. Excess water is exuded in the process of guttation.

Leaves abscise in the fall to allow the plant to overwinter without the potential water loss. Various chemical compounds are reabsorbed from the leaf, and the middle lamella between the parenchyma cells of the petiole and the cells of the stem is dissolved, and the leaf drops off.

Leaves are highly adapted in some plants, for example as needles in cacti to discourage herbivores. Tendrils and bulbs are modified leaves; perhaps the most highly adapted leaves are seen in insectivorous plants.

Lecture Outline

<u>Leaf structure</u>

- Typically thin and flat to increase the surface area to optimize photosynthesis
- Epidermal tissue
    - Cells are typically nonphotosynthetic
    - Cell wall facing outside is thickened
    - May have many trichomes
        - Fuzzy surface may decrease evaporation by creating dead air zone

- Upper epidermis
  - Thicker cuticle to prevent water loss
  - Particularly thick in plants that live in hot, dry areas
- Lower epidermis
  - Stomata are particularly abundant on the lower epidermis
    - Stomates are flanked by guard cells
  - In many plants, stomates are only found here
  - Absent on lower surface of leaves of water lilies

Mesophyll
- Photosynthetic cells between the epidermal layers
- Loose arrangement of cells; air spaces between cells
- Palisade mesophyll
  - Close to the upper epidermis, cells are closely packed
  - Adapted for photosynthesis
  - One to several layers; thicker leaves seen in plants exposed to direct sunlight
- Spongy mesophyll
  - Cells are also photosynthetic, but prime function is diffusion of carbon dioxide
- Vascular bundles (veins) branch throughout the mesophyll
  - Xylem is located on the upper side of the bundle
  - Phloem is on the lower side
- Veins are surrounded by bundle sheath cells
  - Are either parenchyma or sclerenchyma cells
  - May have bundle sheath extensions for support

Monocot and dicot leaves
- Monocots have parallel veins, leaves tend to be narrow
  - Leaves wrap around the stem like a sheath
  - Mesophyll may not be differentiated
  - Guard cells may be shaped like dumb-bells
- Dicots have netted venation, leaves may be broad
  - Leaves attach to the stem via the petiole
  - Mesophyll typically has both palisade and spongy layers
  - Guard cells are shaped like kidney beans

<u>Leaf function</u>

Photosynthesis

- Sunlight is used to fix carbon in the form of sugar
  - Sugars are then transported elsewhere in the plant via the phloem
- Carbon dioxide is the carbon source
- Ultimately, oxygen is given off
  - Both gases diffuse into the leaves through the stomata
- Water is required for photosynthesis
  - Transported from the roots via the xylem

Stomates

- Typically open in the day
  - When cells become turgid, they bend, and produce the pore
- Low carbon dioxide concentrations induce stomates to open, even at night
- Water stress results in stomatal closing
- Mechanism is affected by plant hormones
- Is also under the affect of a biological clock; circadian rhythms
- Mechanism of opening and closing
  - Based on a potassium ion (K^+) mechanism
  - Light triggers an influx of K^+ ions into the guard cells from surrounding epidermal cells
  - Water then passes into the cells by osmosis
  - Increased water pressure deforms the guard cells
  - When dark, stomates close by a reversal of the process
- Believed that stomata are most strongly affected by blue light
  - A yellow pigment is triggered by blue light, starts the process

Transpiration and guttation

- Leaves lose water from leaves, through the stomates
  - Light increases transpiration losses, because stomates open
  - Heat increases transpiration
- Has a cooling function for the leaf due to evaporation
- Also allows plants to concentrate minerals
  - Water exits leaves during transpiration, minerals do not
- Temporary wilting
  - When a plant loses water and wilts
  - Plants can typically recover from this overnight by closing the stomates
- Permanent wilting results in plant death
- Guttation occurs when transpiration rates are low and soil moisture is high
  - Occurs at night
  - Plants become covered with water droplets which were forced out

Leaf abscission

- Leaves of trees in temperate areas are shed before winter
  - Acts to conserve water, since leaves are primary site of transpiration
    - During the winter, the ground is frozen at the surface; water is unavailable to plant roots
  - Metabolism of the plant slows
  - Plant hormones interact
  - Plant reabsorbs sugars, minerals from leaves
  - Chlorophyll is broken down, carotenes become evident
- Abscission zone
  - At the base of the leaf petiole
  - Composed of parenchyma cells, weak because fibers are sparse
  - Enzymes dissolve the middle lamella, leaf falls off
  - Leaf scar remains

Unusual leaves

Spines are modified leaves seen in cacti
- Stems are the primary photosynthetic organ

Tendrils are specialized leaves seen in climbing vines

Bud scales are modified leaves which protect the apical meristem

Bulbs are underground stems with fleshy leaves
- Onions and tulips

Succulents store water in their leaves

Insectivorous plants
- Leaves are adapted as passive traps
  - Pitcher plant, sundews
- Active traps
  - Venus fly-trap, bladderwort

Research and Discussion Topics

• Relate the structure of the palisade and spongy mesophyll layers to the adaptations seen in C_4 and CAM mechanisms for photosynthesis.

• Examine the adaptations seen in aquatic plants (freshwater, and the few marine species) for photosynthesis.

• Investigate the unique mechanism by which the leaves of the Venus fly-trap utilizes the force of osmosis to close the trap.

• Describe various carnivorous plants, such as bladderworts, sundews and pitcher plants. Why do they typically grow in bogs? Why do they eat insects? The same answer explains both questions.

• Look at the example of convergent evolution of the cacti and the euphorbs. What adaptations do they have for living in arid environments?

• What are the leaves of gymnosperms? Where are stomates found in gymnosperm leaves? Why is this adaptive?

Teaching Suggestions

• A very effective illustration of wilting can be done with a little advanced planning. Allow a small plant to wilt, take it to class and water it at the beginning of the lecture. Hopefully, you will see it become turgid during the lecture. If not, you have demonstrated permanent wilting.

Chapter 28. Stems and Roots

Chapter Overview

Stems bear leaves, have vascular tissue for internal transport, and contain meristems to allow growth and differentiation. All stems have primary growth at apical meristems, dicots and gymnosperm stems may also have secondary growth due to growth at lateral meristems. Stems are bounded by epidermis, under which is cortex. Vascular bundles are arranged in a circle in dicots, and deep to them is the pith, composed of parenchyma.

Roots have similar primary tissues, but many unique features such as a root cap and root hairs. Further, their inner tissues are arranged differently than stems. Under the epidermis is the cortex, bounded by the endodermis, which has the unique waterproofed Casparian strip. The pericycle is inside the endodermis and is meristematic. Vascular tissue is in the inside of the root, with xylem arranged in "arms", phloem between the arms.

Water enters the root by moving along the cellulose-rich cell walls, and between cells until it reaches the Casparian strip. Cells of the endodermis regulate passage of water to the inner xylem. Roots store starches, typically in the cells of the cortex. The pericycle is the origin of branch roots.

In both stems and roots, woody plants have secondary growth due to divisions of the vascular cambium and the cork cambium. The variation in growth between the seasons produces annual rings, which may be used to age trees which live in temperate regions.

Water is moved upward in xylem, primarily due to the tension-cohesion mechanism. It is due to the transpiratory loss of water, which pulls water up in the plant, due to cohesive and adhesive properties of water (based on hydrogen bonding). The root pressure mechanism seems to be of lesser importance.

Sugar is translocated primarily from leaves to other plant parts, but may be transported in either direction, from source to sink by the pressure-flow hypothesis. Sugars are actively transported from mesophyll cells to the companion cells, then passively to the sieve-tube cells. Water follows by osmosis, and the resulting higher water pressure in the phloem cells results in movement of the sucrose solution. At the sink, the solution flows from phloem sieve-tube cells to companion cells, then is actively transported out of the companion cells, utilizing the energy from ATP.

Roots actively absorb minerals and ultimately pass the minerals to the xylem for translocation. By studies of hydroponics, it has been determined that sixteen elements are essential for plant growth. Plants have very high requirements for the macronutrients carbon, oxygen, hydrogen (which come from water or the atmosphere) and nitrogen (which comes from the soil). Fertilizers, whether organic or inorganic help the plant meet these nutrient requirements.

Lecture Outline

<u>Stems</u>

- Usually located above ground
- Functions:
  - Bear leaves and reproductive organs
  - Provide internal transport via xylem and phloem
  - Produce new tissue in the apical and lateral meristems
    - Primary growth occurs at apical meristems
    - Secondary growth is seen in woody plants, and occurs at lateral meristems.
- Stem structure (typical dicot)
  - Epidermis on outside
    - Protects the stem
    - Covered by the cuticle
  - Cortex is under the epidermis
  - Vascular bundles are arranged in a circle, xylem to the inside
    - Vascular cambium is between the xylem and phloem
    - Some have a phloem fiber cap
  - Pith in the center of the stem
    - Composed of thin walled parenchyma
    - Pith rays are the areas between vascular bundles
- Monocot stems are similar to dicot stems, with a few exceptions:
  - Vascular bundles are randomly scattered
  - Pith and cortex are not differentiated; simply called ground tissue (parenchyma)
  - Lack vascular cambium
- Function of cortex and ground tissue
  - Photosynthesis
  - Storage
  - Support
- Function of vascular tissues
  - Conduction of materials
  - Support of the plant body, particularly due to fibers

<u>Roots</u>

- Function to anchor the plant
  - Transport water and dissolved minerals to upper part of plant
  - Many are storage organs
  - Some plants have aerial photosynthetic roots
- Structures similar to stems:
  - Epidermis, cortex, vascular bundles, pith
- Unique structures:
  - Root cap
    - Protects the root apical meristem

- Root hairs
  - Extensions of the epidermis which increase the surface area of the root
- Root epidermis does not secrete a cuticle
- Cortex has loosely arranged parenchyma
  - Deep to the cortex is the endodermis
    - Unique cells, have Casparian strip on radial and transverse walls
    - Contains suberin which is hydrophobic
  - Inside the endodermis is the pericycle
    - Pericycle is meristematic
- Inner core is vascular tissue
  - Xylem is arranged in several extensions
  - Phloem is between xylem "arms"
  - Vascular cambium is between xylem and phloem
- Monocot root is similar, but xylem and phloem do not compose the entire center of the root, but are in a circle around the central pith

Function of roots

- Absorption of water is facilitated by the lack of cuticle, presence of root hairs
- Cell walls are made of cellulose, which absorbs water readily
  - Think about cotton (cellulose) towels!
  - Water moves laterally along cell walls and intercellular spaces until it reaches the endodermis
    - Cells of the endodermis control water passage to the xylem
    - Water enters the xylem arms, now moves upward
- Root cortex stores starch, a polymer of glucose
  - Phloem carries dissolved sugars to the roots or from the roots to other parts of the plant
- Intercellular spaces allow for aeration
  - Oxygen diffuses from air spaces of the soil to the root
- The meristematic pericycle is the source of root branches

<u>Woody stems and roots</u>

Dicots and gymnosperms have lateral meristems

- Vascular cambium
  - Cells divide and produce secondary tissues
    - Secondary xylem (wood)
    - Secondary phloem (inner bark)
  - Replacements for primary xylem and phloem
  - When meristematic tissue divides, one daughter cells remains meristematic
  - Cells divide in two planes
  - Cells can be formed inside (secondary xylem) or outside (secondary phloem)

Cork cambium

Produces cork cells and cork parenchyma, the periderm or outer bark

Bark is the replacement for the epidermis

Trees may be aged by rings in wood (trees in temperate areas)

In spring, new cells are large and thin-walled (springwood)

In summer, they are thick-walled and have many fibers (summerwood)

<u>Transport in xylem</u>

Translocation refers to transport in phloem or xylem

Transport in xylem occurs in one direction only

Xylem transports water and dissolved minerals

Move within tracheids and vessel elements (both dead cells)

Transported quickly, up to 2 feet/minute

Water is pulled through the xylem- the tension-cohesion mechanism

Transpiration provides the evaporative pull

Solar radiation then is the ultimate cause

Cohesiveness of water molecules, due to hydrogen bonding

Adhesion to the walls of the xylem cells

The root pressure mechanism describes that water moves from the soil into the xylem by osmosis

Probably a minor mechanism compared to the pull of transpiration

<u>Transport in phloem</u>

Translocation occurs both upwards or downwards

After production by photosynthesis, glucose is converted to sucrose

Translocation in phloem is slower, about 1 inch per minute

Phloem transports sugars from sources to sinks

Pressure-flow hypothesis

Due to a pressure gradient between source and sink

Sucrose moves from mesophyll into companion cells by active transport

Requires ATP

Then moves into sieve-tube member via cytoplasmic connections

Water follows by osmosis

This causes an increase in pressure in the sieve-tubes

The sugar solution is then pushed through the phloem

At the sink, sugar is actively transported from sieve-tube cells to companion cells

Requires ATP

Water follows by osmosis

<u>Root absorption of minerals</u>

Minerals may be in xylem in differing concentration than soil water

Most minerals enter by passing through the plasma membranes of epidermal root cells

- Travel from cell to cell
- Active transport, requires much ATP

Essential elements; 16 known as essential to plants

- 9 are macronutrients; C, H, O, N, P, K, S, Ca, and Mg
  - C, O, H come from water or atmosphere
  - N comes from soil
- 7 are micronutrients; Fe, B, Mn, Cu, Mb, Cl, Zn
- Studied by hydroponics
- Fertilizers replace essential elements
  - Typically N, P and K are limiting
  - Organic fertilizers
    - Advantageous because increases the amount of organic matter, releases nutrients slowly, and pathogenic microorganisms
  - Inorganic fertilizers are manufactured
    - Numbered by proportion of N : P : K

Teaching Suggestions

• Cellulose is an interesting molecule. Cellulose of the cell walls of the root absorbs water like a wick, which allows plants to readily absorb water. Cellulose absorbs water, but it isn't like most hydrophilic substances, which dissolve in water. We make towels of cotton instead of rayon so they can absorb water. Fortunately it doesn't dissolve in water, or we'd never be able to wash our towels!

• This is a good point to discuss the care of house and garden plants. Students have learned why we water them, to keep their cells turgid, now they learn why you don't want to soak your potted plants!

Suggested Readings

Kendler, B.S. and H.G. Koritz. "Using the Allium test to detect environmental pollutants." *American Biology Teacher*. September 1990. 52 (6): 372-375. Use of onion roots to detect environmental contaminants.

Smith, M. "Plant growth-responses to touch – literally a 'hands-on' exercise!" *American Biology Teacher*. February 1991. 53 (2): 111-114. The effects of touching stems on bean plant height.

Chapter 29. Reproduction in Flowering Plants

Chapter Overview

Flowering plants may reproduce both sexually and asexually. Typical methods of asexual reproduction involve modified stems or roots. Various modified underground stems or leaves, such as rhizomes, tubers, bulbs and corms may aid in the propagation of plants. Above-ground stems which are horizontal (runners or stolons) are seen in strawberries. Some leaves can form little plantlets; some roots can form above-ground stems (suckers). Some plants can form embryos without meiosis, known as apomixis.

Flowers are reproductive structures which ultimately produce the male gametophyte; the pollen grain, and the female gametophyte, formed within the ovule. During the process of pollination, pollen is transferred to the stigma. This can be accomplished by animals, wind, water, or wind. Flowers have many adaptations to facilitate pollination. After pollination, the pollen tube grows down into the embryo sac, and the two sperm travel down the tube. During double fertilization, one sperm fertilizes the egg, forming the embryo, and the other fuses with the two polar nuclei, forming endosperm. Endosperm is nutritive material, and may be persistent in the seed, or absorbed by the cotyledons.

The embryo goes through a series of stages to form the mature seed, which is composed of the embryo, food, and the protective seed coat. In angiosperms, seeds are enclosed in the fruit. Fruits may simple, aggregate, multiple or accessory, depending on the floral parts contributing to the fruit, and the number of carpels involved. Seed dispersal is also varied; seeds are dispersed by wind, animals, water and explosive dehiscence.

Lecture Outline

Reproduction in flowering plants

- Many can reproduce sexually and asexually
  - Sexual reproduction involves flowers, seeds and fruits
    - Involves meiosis
    - Genetic variation promoted by independent assortment and crossing over
  - Asexual reproduction involves propagation of vegetative structures
    - Offspring are produced by mitosis
    - Offspring are genetically identical to the parents

Asexual reproduction

- Modified stems may produce entire plants when separated from the parental unit

Rhizomes
- Horizontal underground stem
- May be fleshy for storage
- Resemble roots, but have stem structures like leaves, buds and nodes
- Examples: irises, bamboo, many grasses

Tubers
- Enlarged for food storage
- Example: potatoes; the eyes are lateral buds
  - Potatoes are grown by planting "eyes"

Bulbs
- Short underground stems with fleshy leaves
- Example: lilies, tulips and onions

Corm
- Underground stem which is the storage organ
- Covered with papery scales which are the leaves
- Examples: crocus, gladiolus

Stolons
- Runners; horizontal above-ground stems
- Long internodes
- New plants come from adventitious buds
- Example: strawberry

Plantlets
- Some plants can produce young plants along leaf margins
- Plantlets drop off, produce new plants

Suckers
- Above-ground stems from adventitious buds on roots
- Examples: locust, pear, apple and blackberry

Apomixis
- Production of embryos asexually
- Examples: dandelions, garlic, some grasses, citrus trees

<u>Sexual reproduction of flowering plants</u>

Flower parts
- Sepals
  - Leaflike, protect the floral parts while in the bud stage
  - All of the sepals make up the calyx
  - Some sepals look like petals, e.g. lilies
- Petals
  - Typically colored
  - Collectively called the corolla
  - Important in attracting animal pollinators
- Stamens
  - Male reproductive organs
  - Made of a filament (stalk) and an anther, where pollen is formed
    - Pollen is the male gametophyte
    - Pollen grains contain two sperm cells

- Carpel(s)
  - Female reproductive organs
  - Stigma; where pollen sticks
  - Style; inferior to the stigma
  - Ovary with one or more ovules
    - Ovules contain the female gametophytes, with the eggs
  - The pistil is the entire carpel or fused carpels

Female reproductive structures
- Cells within the ovules undergo meiosis and produce 4 haploid megaspores
- One forms the female gametophyte (embryo sac)
  - Consists of 8 haploid nuclei, 1 egg and 2 polar nuclei

Male reproductive structures
- Cells within the anther undergo meiosis and produce 4 haploid microspores
- Microspores form the male gametophyte (pollen grain)
- Pollen grain contains 2 sperm cells; both participate in fertilization

Fertilization
- Involves transfer of pollen to the female structures of the flower
- Various mechanisms
  - Petals, nectar, pollen, scent attract pollinators
    - Bees are attracted to blue or yellow
      - Also ultraviolet markings (bee's purple)
    - Flies are attracted to strong scents
    - Hummingbirds are attracted to red colors, nectar
    - Bats are attracted to white flowers, strong scents
- Coevolution between flowers and pollinators
  - Example: orchids which smell like sexually receptive wasps
- Wind-dispersed pollen is produced in inconspicuous flowers
  - Typically produce much pollen

Fertilization
- Pollen tube grows down style and into embryo sac
- Two sperm pass down tube
- Double fertilization
  - One sperm fuses with egg
    - Produces the embryo
  - Other sperm fuses with two polar nuclei
    - Produces the endosperm

Embryo development
- Embryo undergoes mitosis to form 2 cells
  - Top cell becomes the embryo
  - Bottom cell becomes the suspensor
    - Aids in nutrient uptake from endosperm
    - Anchors the embryo
- Embryonic cell divides to form proembryo, then the globular embryo
  - Dicot embryos have 2 cotyledons, called the heart stage
  - Monocot embryos have 1 cotyledon; is more cylindrical

- Develops to form the torpedo stage; enlarges, crushes the suspensor
- Mature embryo consists of embryonic root and shoot and cotyledons
    - Monocots have one cotyledon
    - Dicots have two cotyledons
- Cotyledons may absorb the nutrients in the endosperm; are nutritive
- Mature seed consists of:
    - Embryonic plant
    - Food stored in cotyledons or endosperm
    - Seed coat
- Seeds are enclosed within the fruit (ripened ovary)

<u>Fruits</u>

- Ovule develops into the seed
- Ovary develops into the fruit
    - May contain one or more seeds
- Simple fruits
    - Developed from a single ovary from one flower
    - May be fleshy
        - Berries, which have soft tissues
            - Examples: tomatoes and grapes
        - Drupes which have a hard pit around the seed
            - Examples: peaches and avocados
    - May be dry
        - Dehiscent fruits (split open at maturity)
            - Examples: milkweed pods, peas
        - Indehiscent fruits (do not split)
            - Example: grains
- Aggregate fruits
    - Developed from a single flower with many carpels
        - Examples: blackberries and raspberries
- Multiple fruits
    - Developed from many flowers which fuse
        - Example: pineapple
- Accessory fruit
    - Develop from ovaries and other plant parts
        - Example: strawberry (the receptacle)
            - Apples (the floral tubes, inferior to the flower)

<u>Seed dispersal</u>

- By wind, water, animal or explosive dehiscence
- Wind dispersal
    - Light fruits
    - Examples: dandelions and milkweeds
- Dispersal by animals
    - Some seeds have hooks that catch mammalian fur
    - Fruit that is fleshy and tasty are dispersed by animals that eat them
    - Some animals store and bury fruits

Dispersal by water
Example: coconut

Research and Discussion Topics

• Describe the mechanisms and characteristics of plants which are pollinated by flies, bees, moths, hummingbirds and butterflies. What different characteristics differentiate flowers pollinated by different types of pollinators?

• Some plants are wind-pollinated. What characteristics might you expect in wind-pollinated flowers like the flowers of grass plants?

• Living in freshwater poses a unique problem for plants to disperse their pollen. What adaptations do freshwater plants have for pollination?

• Relatively few higher plants are marine. How do they reproduce?

Teaching Suggestions

• It's fun to bring in fruits of various types and discuss what floral parts they represent. For example, what are those little "things" on the apple, at the end opposite the stem (answer: anthers and stigmas)? What is that little thing on a pea pod, opposite the stem? (answer: the stigma) Students enjoy learning that many things they call vegetables are really fruits. We also discuss what vegetables are really vegetables, like lettuce, onions and broccoli. Relate what plant parts those vegetables are (stems, bulbs etc.).

Suggested Readings

Handel, S.N. and A.J. Beattie. "Seed dispersal by ants." *Scientific American*. August 1990. 76-83.

Meyerowitz, E.M. "The genetics of flower development." *Scientific American*. November 1994. 56-65. Genetic studies of *Arabidopsis* flowers.

Tatina, R. and K. Hohn. "A technique for staining pollen nuclei." *American Biology Teacher*. March 1994. 56 (3): 174-175. A "how-to-do-it" column to demonstrate stained pollen tubes to students.

Chapter 30. Regulation of Plant Growth and Development

Chapter Overview

Plants are affected by various environmental factors which trigger flowering, movements etc. Photoperiodism is marked in plants, which may be described as short-day plants, long-day plants or day-neutral plants. Photoperiodism is strongly affected by the conversion of P_r to P_{fr} during the day. Short-day and long-day plants respond differently to these phytochromes. Temperature also affects the timing of reproduction.

Germination is also affected by environmental factors, such as water, oxygen, temperature and light. Dicot and monocot seedlings are different, but both protect the delicate apical meristem as the shoot pushes through the soil.

Circadian rhythms affect stomatal opening and closing, and sleep movements. Other movements of plant, such as the closing of leaflets of the sensitive plant and solar tracking are caused by ionic and osmotic changes. Tropisms include geotropism, gravitropism and thigmotropism.

Plant hormones have various and complex effects, as do animal hormones. Auxins primarily act to cause cell elongation, and to promote apical dominance. Seeds produce auxins to stimulate fruit development. Gibberellins cause stem elongation and stimulate flowering, particularly in long-day plants. Cytokinins stimulate cell division and differentiation, and act in an antagonistic manner to auxins, as cytokinins stimulate growth of lateral buds. Ethylene is the only gaseous hormone, and inhibits cell elongation and promotes seed germination and fruit ripening. Abscisic acid does not cause abscission, but promotes changes such as stomatal closing during water stress. Abscission, rather, is caused by the interactions between ethylene, auxin and cytokinins.

Lecture Outline

<u>Environmental clues trigger flowering in plants</u>

- Photoperiodism
    - Responses of plant to lengths of light and darkness
    - Flowering is triggered by photoperiodism
    - Short day plants
        - Require long uninterrupted periods of darkness
        - Examples: chrysanthemums and poinsettias
        - Flower in late summer
    - Long day plants
        - Require shorter periods of darkness
        - Examples: lettuce, clover
        - Flower in early summer

Day neutral plants
Respond to other stimuli for flowering
Examples: tomatoes, pansies
Phytochromes trigger photoperiodism
Phytochromes are the photoreceptor in plants
Two forms:
P_r is the red absorbing phytochrome
P_{fr} is the far-red absorbing phytochrome
Less stable, reverts to P_r in the dark
Is the form which triggers responses like flowering
Sunlight contains more red than far-red light, so it triggers the production of P_r
In short-day plants, P_{fr} inhibits flowering
During long nights, P_r changes into P_{fr}, allows flowering
In long-day plants, P_{fr} induces flowering
During long days, the plants produce P_{fr}, some reverts at night, but not all
Not known why the plants respond so differently to P_r and P_{fr}
Temperature also affects reproduction
Some plants need a period of vernalization (cold temperatures) before they will flower
Others require seeds or seedlings to be exposed to cold for flowering

<u>Factors affecting germination and early growth</u>
Germination is often affected by environmental factors
Seeds don't germinate unless water is present
Optimal so seeds germination only when sufficient water surrounds them
Seeds don't germinate unless oxygen is present (need it for aerobic respiration to produce ATP)
Seeds need a particular optimal temperature
Some seeds need light, to convert P_r to P_{fr}
Often a prolonged cold period is needed for germination
Optimal so seeds germinate in spring
May be genetic factors as well
Early growth patterns
The embryonic root is the first to emerge from the seed
Apical meristem of root tip is protected by root cap
Dicot stem meristems are protected by the hooked stem
Monocot stem meristems have a protective coleoptile
The coleoptile pushes through the dirt, leaves grow up through it

<u>Plant responses</u>
Circadian rhythms
May be based on phytochromes
Example: closing and opening of stomata

Sleep movements in plants

Movement of plant parts

Mimosa, the sensitive plant

Touching it causes a signal to travel to the pulvinus, at the bottom of the petiole, and a change in turgor causes the leaves to move

Solar tracking

Often have pulvini

Example: sunflowers

Tropisms are directional responses

May be positive or negative

Phototropism

Shoots grow towards light

Gravitropism

Shoot tips grow against the force of gravity

Thigmotropism

Shoot tips curl around solid objects

<u>Hormones mediate many plant responses</u>

Like animal hormones, plant hormones are produced in one plant part and affect other parts

Produced in very small amounts

Have overlapping effects

May be stimulatory in one concentration, inhibitory in different concentrations

Five classes of plant hormones

<u>Auxins</u>

Darwin showed first evidence for auxins

Produced in coleoptiles, produce positive phototropism

Now known to be produced in stems and coleoptiles

A group of related chemicals; primary auxin is indoleacetic acid (IAA)

Promote cell elongation by changing cell walls

May be involved in gravitropism and thigmotropism as well

Other effects:

Produced in apical meristems of some plants

Causes apical dominance; inhibits lateral growth

Pinch off apical bud, get a "bushier" plant

Auxin produced by seeds stimulates development of fruit

If applied to flowers in which fertilization hasn't occurred, the ovary develops into a seedless fruit.

Implicated in leaf abscission; auxin levels decline in the fall

Synthetic auxins have been made

One is used to stimulate root development by nurserymen

Another, 2,4-D is a herbicide (a component of Agent Orange) used during the Vietnam war

Gibberellins

First knowledge of gibberellins came from fungi which also produce them
Causes stem elongation in many plants
  Gardeners know the effects of bolting
Stimulate flowering, can cause larger fruits to be produced (grapes)
Also involved in the germination of seeds

Cytokinins

First identified in coconut milk and herring sperm
Induces cell division (cytokinesis)
A necessary ingredient in plant tissue culture
Interacts with auxins in apical dominance
  Cytokinins promotes lateral bud development
Delays senescence
  Commercially added to cut flowers to delay browning
May be involved in leaf abscission; levels decline in the fall

Ethylene

A gaseous hormone
Inhibits cell elongation
Promotes germination of seeds
Triggers ripening
  Used commercially to make fruit all ripen synchronously
  Examples: tomatoes and bananas
Involved in leaf abscission (action opposite of auxins and cytokinins)

Abscisic acid

Primarily involved in inducing dormancy, not abscission
Plant "stress hormone"
Produced in response to water stress, triggers stomatal closing
Causes dormancy in seeds

Research and Discussion Topics

• How are plant hormones, like giberellin and auxins, used in agriculture or the nursery business?

• Compare and contrast the production and action of plant and animal hormones (see chapter 41).

Teaching Suggestions

• Discuss what effects pruning a plant has (makes it "bushier"). Relate this to apical dominance and the removal of auxins. Should you prune all plants? For example, do you pinch back your tomato or zucchini plants?

Chapter 31. Animal Tissues, Organs and Organ Systems

Chapter Overview

Multicellular animals are organized in a hierarchy; cells into tissue, tissues into organs, and organs into organ systems. This allows for specialization of function at all three levels. Tissues may be classified in functional groups: epithelial, connective, muscular and nervous. Epithelial tissue functions in protection, secretion, absorption and sensation. Epithelial cells are specialized for these functions by different arrangements (stratified or simple), and shapes (squamous, cuboidal or columnar).

Connective tissues support and cushion other structures of the body. It is characterized by much extracellular substance (matrix) and fibers. Different types of cells and matrix characterize the wide variety of connective tissues; bone, blood, dense connective and adipose tissues.

Muscular tissue is specialized to contract, and is classified as skeletal, which moves bones; cardiac, which composes the walls of the heart, and smooth, which composes the walls of many internal organs. Finally, nervous tissue is specialized to receive stimuli and transmit nervous impulses.

Coordinated tissues form organs, and coordinated organs form organ system. Organ systems work together to maintain a dynamic equilibrium. This equilibrium is regulated by negative feedback mechanisms, and less frequently, by positive feedback mechanisms.

Lecture Outline

Organization of animals

- Multicellular organization allows specializations
- Cells are organized into tissues
    - Tissues are organized into organs
        - Organs are organized into organ systems
- Variations in numbers of cells results in variations in size of organisms

Tissue types

- Tissues may be defined as cells adapted for particular functions
    - Tissues in animals are epithelial, connective, muscular and nervous

Epithelial tissues

- Covers and lines surfaces of the body
- Cells fit tightly together
- Basement membrane (fibrous) anchors it to tissue underneath

Functions:
- Protection, as seen in the skin
  - Subject to constant wear and tear
  - Characterized by rapid mitosis
- Absorption, as seen in the small intestine
- Secretion, as seen in the glands lining the stomach
- Sensation, as in the olfactory epithelium

May be simple (one cell thick) or stratified (layered)
- Pseudostratified epithelium appears to be stratified, but all cells are in contact with the basement membrane

Cell shape:
- Squamous- plate like
- Cuboidal- roughly cube shaped
- Columnar- elongate

Free edge may have cilia or microvilli

Epithelial glands
- Exocrine glands secrete their product via a duct
  - Example: sweat glands
- Endocrine glands secrete hormones into the blood
  - Example: thyroid gland

Connective tissue

Supports and protects the body

Examples: bone, blood, fat

Structure:
- Much intercellular matrix with fibers
- Matrix is composed of polysaccharides, salts or water
- Fibroblasts produce the matrix and fibers
  - Fibers are collagenous (most abundant), elastic and reticular
    - Collagen fibers add strength
    - Elastic fibers add elasticity
    - Reticular fibers provide delicate support

Loose connective tissue
- Most widely distributed; found all over the body
- Fills spaces between other tissues, stores fluids and salts
- Covers nerves, bones, muscles
- Attaches skin to underlying muscle

Dense connective tissue
- Collagen fibers predominate
- Dense regular connective tissue- fibers all run parallel
  - Strength in one direction, like ligaments and tendons
- Dense irregular connective tissue- fibers run in all directions
  - Strength in all directions, as found in the dermis of the skin

Adipose tissue
- Cushions internal organs
- Found in subcutaneous layer
- Cells have a huge fat vacuole; organelles are pushed to the periphery

Cartilaginous tissue
- Forms embryonic skeleton
- Retained in the adult in the external ear, nose, ends of bones
- Cells are chondrocytes, found in pairs in lacunae
- Tissue is avascular, lacks nerves

Bone
- Forms typical vertebrate skeleton
- Cells are osteocytes, also in lacunae
- Osteons are the functional unit of the bone
  - Osteocytes are arranged in layers, lamellae
  - Central Haversian canals carry blood vessels, nerves

Blood and lymph are also connective tissues
- Consists of cells (red and white blood cells, platelets) in a fluid matrix
- Function in circulation of gases, nutrients etc.

<u>Muscular tissue</u>

Functions in contraction
- Cells are known as fibers
- Three types: skeletal, cardiac and smooth

Skeletal muscle
- Attached to bones
- Voluntary contraction
- Is striated; has visible cross markings as a result of the pattern of overlap of microtubules
- Is multinucleate

Cardiac muscle
- Makes up the muscle of the heart
- Striated, but not under voluntary control
- Also multinucleate
- Fibers branch
- Intercalated discs where fibers join

Smooth muscle
- Involuntary, nonstriated, not multinucleate
- Forms walls of many internal organs

<u>Nervous tissue</u>

Receives stimuli and transmits nerve impulses

Composes nerves, spinal cord and nerve

Neurons are nerve cells, supported by nonneural glial cells
- Neurons are composed of the cell body, from which dendrites and axons extend

<u>Hierarchical levels of organization</u>

Organs are composed of coordinated tissues
- Have a specialized function

Organ systems are composed of coordinated organs and tissues
- Typically recognize ten organ systems

Organ systems work together to maintain homeostasis
- Homeostasis is a dynamic equilibrium
- A basic concept of physiology
- Feedback systems help maintain homeostasis
  - Negative feedback systems (most common)
  - Positive feedback systems

Research and Discussion Topics

• In evolutionary terms, when did each of the major systems appear? For example, what animal is considered to have a nervous system? A digestive system?

• What plant tissues are analogous to the four types of animal tissues? What similarities exist in structure and function?

Teaching Suggestions

• To begin the units on vertebrate systems, I ask students to try to list the systems. I write them on the board in the order in which they call them out. The last few are always tough for them to come up with, particularly the lymphatic. I point out that the order in which they think about them is a reflection of their familiarity with them (digestive, for example is always one of the first ones mentioned).

Suggested Readings

Bishop, J.M. "The molecular genetics of cancer." *Science*. 16 January 1987. 305-236. Viruses, chromosomes and oncogenes affect neoplastic cells.

Moss, R. "Cancer education: relieving the silence and fear." *American Biology Teacher*. Nov/Dec. 1992. 54 (8): 458-462. Describing the most common cancers, and methods of screening and detection.

An excellent source of information on human health issues is the University of California at Berkeley Wellness letter. It is a short newsletter which is just packed with current, interesting nuggets of information. For example, in the recent issue, there was information about the prostate test, kidney stones, lowfat cookies, sweating, and the SPF of clothing! Subscriptions are $24/year; write to The subscription Dept, Box 420148, Palm Coast, Florida.

Chapter 32. Skin, Bones, and Muscle: Protection, Support, and Locomotion

Chapter Overview

The epithelium of animals may be adapted for protection or secretion. The vertebrate skin is particularly adaptive, and may form scales, nails or claws, hair or feathers. Many glands are epidermal in origin, such as mammary glands and sebaceous glands. The epidermis of mammals is protective due to accumulation of keratin, and the dermis is strong, and contains many epidermal structures such as hair follicles and sweat glands.

Simple hydrostatic skeletons allow soft-bodied invertebrates to move, and reach the height of specialization in segmented worms. Exoskeletons act to protect invertebrates like molluscs and arthropods, as well as providing a site for muscle attachment. Unlike exoskeletons, endoskeletons are living and can grow, and also protect and allow movement.

In vertebrates, the skeleton is composed of the axial and appendicular skeleton. Typical long bones are composed of hard compact bone, and lightweight spongy bone. On a microscopic level, bones are composed of tiny functional units, the osteons. Osteons are produced by osteoblasts, which lay down the mineral salts and the collagen. Osteoclasts are important in bone remodeling. Varying types of joints connect bones; the synovial joint is the most complex, and allows the greatest degree of movement.

The sole function of muscular tissue is to contract, causing movement. Muscle cells (fibers) are very complex structures, with specializations of the plasma membrane and endoplasmic reticulum functioning in contraction. The bulk of the cell is made of actin and myosin filaments. Muscle contraction begins when the nerve impulse from a motor neuron first depolarizes the plasma membrane, and then the T-tubules. Calcium is released, which results in the attachment of myosin heads to the binding sites of the actin filaments. As the crossbridges flex, and reattach, the filaments slide past each other. The fuel for this process ultimately comes from glycogen, the storage form of sugar. Creatine phosphate stores energy, which is transferred to ATP as needed. When oxygen demand exceeds supply, cells switch to lactic acid fermentation, which results in muscle fatigue.

The contraction of individual muscle fibers results in contraction of the entire muscle, which moves parts of the body by pulling on bones. Skeletal muscles are arranged in antagonistic pairs. Smooth muscle contracts more slowly than cardiac or skeletal muscle. Skeletal muscles may contract singly; a simple twitch, or in a sustained contraction, called tetanus. Fast-twitch fibers are able to contract quickly, slow-twitch fibers contract more slowly. The proportion of these two types of fibers is genetically determined, but may be changed by athletic training.

Lecture Outline

Epithelial covering of the body of animals

- Invertebrate epithelia
  - Sensory function
  - May have tough cuticle
  - May be glandular, producing lubricants, poisons, or odors
    - Silk glands of spiders, butterflies and moths are epithelial glands
- Vertebrate skin
  - Amphibian skin is glandular and moist
    - A few have poison glands
  - May form scales, as in fish, reptiles and a few mammals
  - Form feathers in birds
  - Specialized structures in mammals:
    - Fingernails, hair
    - Various glands
      - Oil glands
      - Mammary glands are specialized epithelial glands
    - Sensory receptors
- Epidermis
  - Outer layer
  - Multiple layers; stratum basale is deepest, stratum corneum is outer
  - Avascular
  - Cells become keratinized by accumulating keratin
    - Functions in waterproofing
    - Cells die, slough off, constantly replaced
- Dermis
  - Dense irregular connective tissue
    - Collagen adds strength
  - Sweat glands (epidermal origin) are embedded in the dermis
  - Hair follicles (also epidermal origin) extend into dermis
  - Is vascular, has many nerve endings
  - Subcutaneous fat below dermis
    - Important in insulation, cushioning

Skeletal systems

- Simplest animals have a hydrostatic skeleton
- Exoskeletons are external skeletons, such as seen in arthropods
- Endoskeletons are internal skeletons, seen in echinoderms and vertebrates

Hydrostatic skeletons

- Muscles contract around a fluid-filled cylinder
  - Circular muscles contract, cylinder lengthens
  - Longitudinal muscles contract, cylinder shortens
- Typically, movements are not precise
- Annelids have separate segments; each has a functional unit of the hydrostatic skeleton

Molluscs use a hydrostatic skeleton to extend their foot to dig
Echinoderms have a highly adapted water vascular system
Mammalian penises work on a similar principle

<u>External skeletons</u>

- Molluscs and arthropods have nonliving exoskeletons
  - Functions:
    - Protection of the body
    - Sites for muscle attachment
- Exoskeleton cannot grow, so animal must molt

<u>Endoskeletons</u>

- Are living tissue; can grow as the animal grows
- In echinoderms; plates covered by a thin epidermis
- Vertebrate endoskeleton functions
  - Protection of internal organs (cranium and rib cage)
  - Attachment for muscles
- Axial skeleton:
  - Skull; cranial and facial bones
  - Vertebral column
    - Cervical, thoracic, lumbar vertebrae, and the sacrum and coccyx
  - Ribs attach ventrally to sternum, dorsally to vertebrae
    - True ribs are attached to sternum via individual cartilages
    - False ribs are attached to last true rib's cartilage
    - Floating ribs have no ventral attachment point
  - Sternum
- Appendicular skeleton:
  - Pectoral girdle
    - 2 clavicles
    - 2 scapulae
  - Pelvic girdle
    - Less flexible than pectoral girdle
    - Composed of three fused hipbones
  - Appendages
    - Typically terminates in 5 digits
    - Modified in some mammals, like horses,and other ungulates
    - Great apes and humans have an opposable thumb

<u>Structure of bones</u>

- Bones act as levers, moved by muscles
- Periosteum is osteogenic; covers the bone
- Diaphysis is the shaft of the bone
- Epiphyses are the ends of the bone
  - Epiphyseal plate is the site of growth between diaphysis and epiphysis
  - Seen only in growing bones
- Bone marrow fills the central cavity
  - Red marrow functions in producing blood cells

- Yellow marrow, seen in adults, is composed of fat
- Compact bone composes the diaphysis, covers the epiphyses
  - Characterized by organized osteons (Haversian systems)
  - Osteocytes are in lacunae
  - Blood vessels travel in Haversian canals
- Spongy bone makes up the interior of the epiphyses
  - Strong, yet light
- Bone growth and repair
  - Bones develop in the fetus in two ways
    - Endochondral bone development; cartilage replacement
    - Membranous bone development seen in skull bones, vertebrae
  - Osteoblasts form new bone tissue
    - When osteoblasts become trapped in new matrix, now called osteocytes
  - Osteoclasts resorb bone
    - Important in reshaping bone during growth
- Joints are found between bones
  - May be immovable, such as sutures of the skull
  - Slightly movable joints are cartilaginous, such as intervertebral joints
  - Freely movable joints (synovial joints)
    - Enclosed by a joint capsule, filled with synovial fluid
    - Held together by ligaments
  - Osteoarthritis is a degenerative joint disorder
  - Rheumatoid arthritis is an autoimmune condition resulting in inflammation of the joints

<u>Muscular tissue</u>

- Muscle action based on contractile proteins
  - Actin and myosin result in contraction of muscles
- Invertebrate musculature
  - Cnidarians have simple contractile cells
  - True muscular tissue first seen in flatworms
  - Bivalves have smooth and striated muscles
    - Smooth muscles can contract to hold the shell shut for days!
  - Arthropod muscles are typically striated
- Vertebrate muscle microanatomy
  - Fibers are arranged in bundles, wrapped in connective tissue
  - Fibers are long, multinucleate
    - Plasma membrane has multiple invaginations called T tubules
    - Cytoplasm is referred to as sarcoplasm
    - ER is sarcoplasmic reticulum
  - Myofibrils
    - Thick filaments - myosin
      - Have arms with heads
      - May bind to the adjacent actin filament
    - Thin filaments - actin and regulatory proteins
    - Filaments alternate and overlap at the ends

Arranged in functional units called sarcomeres
Sarcomeres are bounded by the Z lines
(Z ends the sarcomere; Z ends the alphabet)

Muscle contraction

- Results from shortening of the sarcomeres
  - At rest, actin's regulatory proteins prevent myosin heads from binding
- Process:
  - Acetylcholine (neurotransmitter) from motor neuron is released
  - Acetylcholine stimulates receptors on muscle fiber
  - Plasma membrane depolarizes
  - Action potential spreads through T tubules
    - Calcium is released into cytoplasm
  - Calcium stimulates regulatory proteins to uncover binding sites
  - Using energy from ATP, myosin heads form cross bridges with actin
  - Cross bridges flex and reattach, pulling filaments past one another
    - Muscle contracts as filaments continue this process
- Muscle tone is a result of contraction of some muscle fibers
- ATP provides the energy for contraction
  - Energy required to connect to, and break at active sites
  - Rigor mortis is due to depletion of ATP
- Creatine phosphate stores energy
  - Transferred to ATP as needed
- Glycogen is a polysaccharide; the storage form of glucose
- Cells perform aerobic respiration as long as oxygen is present
  - Fermentation pathways when oxygen is depleted
    - Produces less ATP
  - Oxygen debt is due to buildup of lactic acid, glycogen depletion
- Skeletal muscles connect to bones
  - Connected to bones via tendons
- Skeletal muscles work in antagonistic pairs
  - Muscles producing a certain action is the agonist
  - Muscle producing opposite action is antagonist
  - Muscles work together to flex and extend

Responses of the three muscle tissue types

- Smooth muscle
  - May contract in response to stretching
  - Slow, strong contraction
- Cardiac muscle
  - Quick contraction, does not tire
- Skeletal muscle
  - Single quick contraction; simple twitch
  - Smooth sustained contraction is tetanus
  - Fast-twitch fibers contract more quickly
    - Many mitochondria
    - Highly vascularized

Would predominate in a sprinter
Slow-twitch fibers contract more slowly
Much myoglobin
Would predominate in a marathon runner
Proportions are variable, genetically determined, may be altered by training

Research and Discussion Topics

• Do all animals have the three types of muscular tissue? Research the development of the three types, and list in what animals they appeared first.

• Discuss various joint disorders, including osteoarthritis and rheumatoid arthritis. What are their causes and treatments?

• Osteoporosis is a "hot" topic. Discuss the various proposed causes and treatments for osteoporosis. Research the involvement of calcium, vitamins, and hormones. Hormone replacement therapy is a controversial therapy suggested for post-menopausal women to avoid osteoporotic bone loss. Discuss pros and cons of HRT.

• Investigate modern techniques for joint surgeries and replacements. The knee is the largest joint and the most highly stressed joint in the human body, and advances in knee surgeries are fascinating.

Teaching Suggestions

• Students seem to be fascinated between "dark meat" and "white meat" of a bird. We discuss why the pectoral muscles, which function in flight, are white, while the leg muscles, which function in standing (sustained contraction) are dark. We tend to eat frog's legs, which are primarily white fibers, because they "taste like chicken." Same thing with rattlesnakes. We don't happen to eat toad legs, however, because they would taste "gamey," because of the predominance of red fibers.

Suggested Readings

Anastasiou, C.J. "The sun and our skin: an update for biology teachers. *American Biology Teacher*. Marcy 1991. 53 (3): 137-145. A description of the effects of sun on the skin, particularly cancerous growths, and the importance of prevention.

Various authors; 5 articles on vertebrate locomotion. *Bioscience*. 30 (11): 764-798. Experiment on the interplay between skeletons and musculature.

Chapter 33. Responsiveness: Neural Control

Chapter Overview

A nervous system allows an animal to quickly respond to changes in the internal or external environment. The functions of a nervous system involve stimulus reception, transmission, integration and response by a muscle or gland. The primary center of integration is the central nervous system.

Neurons are the cells which carry the nervous impulses, and are supported by glial cells. Neurons are composed of a cell body, numerous short, branching extensions which carry impulses towards the cell body, known as dendrites, and axons, which are elongate, typically single extensions which carry nervous impulses to neighboring cells. Axons in the peripheral nervous system are invested with a myelin sheath, produced by Schwann cells, which insulate the axon, speed nervous impulse transmission, and allow for regeneration, if the axon is cut.

Nerves transmit nervous impulses in the form of an action potential, which is the depolarization and subsequent repolarization of the plasma membrane of the cell. The resting potential (-70 mV) is maintained primarily by the action of sodium-potassium pumps which concentrate sodium ions outside the cell, and potassium ions inside the cell. When a cell is stimulated, it depolarizes. If the depolarization is sufficient, an action potential will occur. This wave of depolarization will travel down the membrane in a unidirectional manner, due to the absolute refractory period. In myelinated neurons, the depolarization jumps from node to node.

Ultimately, the nervous impulse must be passed from cell to cell. It may be passed directly, via gap junctions in electrical synapses. More commonly, chemical synapses utilizing neurotransmitters bridge the synaptic cleft. As the action potential reaches the synaptic knobs, influx of calcium causes vesicles with neurotransmitters move to and fuse with the plasma membrane. The expelled neurotransmitter stimulates receptors in the postsynaptic cell to propagate an action potential. Some neurotransmitters are excitatory, others are inhibitory. Acetylcholine and norepinephrine are two common neurotransmitters.

Neurons synapse with many other neurons; these connections allow the cell to calculate the sum of all connections, which is a tabulation of the local EPSP's and the IPSP's. Since most cell bodies are located in the CNS, it is the primary center of integration.

Reflex arcs show the neural pathway of simple responses. Withdrawal reflexes are polysynaptic, and a sensory neuron synapses with an association neuron in the CNS, which sends a message directly to the muscle or gland via the motor neuron. Complex neural pathways involve convergence, divergence and reverberating circuits.

Lecture Outline

Function of the nervous system

- Response to internal or external environment
  - Respond to stimuli
    - Rapid response (endocrine system provides a slow response)
  - Transmit nervous impulse to the central nervous system
    - Sensory neurons
  - Integration
    - Processing of nervous signal
  - Response
    - Motor neuron travels to a muscle or gland

Neuron anatomy

- Neurons are nerve cells
  - Neurons are supported by non-neural glial cells (neuroglia)
    - Some neuroglia wrap neuron cell bodies and their cellular extensions
    - Others are phagocytic, or line cavities in the CNS
- Neurons are specialized in reception and transmission of nervous impulses
  - Cell body contains most of the cytoplasm, nucleus and other organelles
  - Dendrites are short, branched extension
    - Adapted to receive nervous impulses
    - Pass nervous impulses to cell body
  - Axons are long extensions
    - Adapted to carry nerve impulses from the cell body to another neuron or muscle
    - Axons end in axon terminals with synaptic knobs
    - Large peripheral axons have a sheath made by Schwann cells
      - Schwann cells wrap around axons
      - Myelin sheath is formed by wrappings of plasma membrane
      - Outer portion of the sheath is the Schwann cell's plasma membrane
      - Myelin is a lipid-rich substance, acts as an insulator
        - Speeds transmission of nerve impulses
      - Gaps between Schwann cells are nodes of Ranvier
      - Sheath allows regeneration of damaged axons
    - Smaller peripheral axons are not myelinated
    - Axons in the CNS are wrapped by other glial cells
    - Multiple sclerosis is an autoimmune disease
      - Myelin deteriorates, interferes with nerve impulse transmission
- Nerves are axons wrapped in connective tissue
  - Cell bodies of these axons may reside in ganglia

The nervous impulse

- Resting potential
  - Membrane is polarized; more negative inside; -70 mV
  - Measured with tiny electrodes
  - Maintained by sodium-potassium pumps
    - Actively transport sodium out and potassium in
  - Ions also cross by facilitated diffusion
  - Negatively charged proteins and organic phosphates also contribute
- The action potential
  - Stimuli alter the permeability of the membrane to sodium
  - Threshold:
    - Depolarization over +15 mV (change of at least -55 mV)
  - Once the threshold is reached, voltage-activated ion channels open
    - Sodium rushes into cell
    - Inactivating gates close the channels
    - K^+ channels also open, close more slowly
  - Polarity of the membrane is reversed; to +35 mV
  - Rise and fall is known as the spike
  - Action potential spreads down the axon; the wave of depolarization
  - Membrane repolarizes as sodium gates close
  - Refractory periods
    - Absolute refractory period- during depolarization
      - Cannot transmit another impulse at all
    - Relative refractory period- during repolarization
      - Cannot transmit another impulse unless it is stronger than the first
- Saltatory conduction
  - Action potentials jump from node to node of Ranvier
  - Faster transmission
  - Requires less energy to repolarize
- All-or-none law
  - Action potential either happens or it doesn't

Synapses

- The synapse is the gap between the neurons, or the neuron and an effector
  - Presynaptic neuron
  - Postsynaptic neuron
- Types of synapses
  - Electrical synapses
    - Impulse directly passes from cell to cell via gap junctions
    - Occurs very rapidly
  - Chemical synapses
    - Cell to cell over the synaptic cleft
      - Neurotransmitters bridge the gap
    - Most common type of synapse
    - Neurotransmitters are produced in the synaptic knobs, stored in vesicles

Action potentials cause calcium to pass into the axon
Calcium causes vesicles to fuse with the PM, release contents
Neurotransmitters affect the postsynaptic cell
May trigger receptors to propagate an action potential
May cause sodium gates to open
Excitatory synapse; causes postsynaptic cell to be closer to firing
May cause potassium channels to open
Inhibitory synapse; causes postsynaptic cell to be less likely to fire
Subsequently, neurotransmitters are broken down by enzymes or reabsorbed into vesicles

Neurotransmitters and related chemicals
Over 60 described
Acetylcholine released by motor neurons, some neurons in CNS, autonomic system
Called cholinergic neurons
Excitatory effect on skeletal muscle
Inhibitory effect on cardiac muscle
Cholinesterase breaks down acetylcholine in the synapse
Norepinephrine released by some neurons in CNS, autonomic system
Called adrinergic neurons
Catecholamines (biogenic amines) include norepinephrine, epinephrine and dopamine
Affect mood, levels associated with depression
Degraded by MAO (monoamine oxidase)

<u>Nerve fibers</u>

Transmission from axon of presynaptic cell to dendrite or cell body of postsynaptic cell
Speed increases with increasing diameter of axon
Myelination increases speed of conduction
Increasing distance between nodes increases speed

<u>Integration</u>

Every neuron synapses with hundreds of neurons
Some synapses are excitatory, others inhibitory
EPSP (excitatory postsynaptic potential) brings neuron closer to firing
IPSP (inhibitory postsynaptic potential) prevents neuron from firing
May cancel each other out; are local responses
"Chemical tabulations"
Permits a wide range of responses in the nervous system
Integration happens at a cellular level; mostly in the CNS

Reflex arcs

An example of a simple neural pathway

Relatively fixed pattern, but some can be learned or modified

Breathing is a simple example of an automatic reflex

Withdrawal reflexes

Polysynaptic; includes three neurons

Sensory neuron sends neural message to the spinal cord

Association neuron synapses

Association neuron transmits message to motor neuron

Motor neuron stimulates muscles to withdraw

Association neuron also sends messages to the conscious portion of the brain

Complex neural pathways

Convergence

Many presynaptic neurons synapse with one postsynaptic neuron

Divergence

One presynaptic neuron synapses with many postsynaptic neurons

Reverberating circuits

Neuron synapses with association neuron in a complex way so that more impulses are generated again and again

Important in rhythmic reflexes like breathing

Research and Discussion Topics

- Investigate various reflexes, such as the patellar, corneal and Babinski's reflexes. Describe the neural circuits that result in these reflexes.

- Describe the mechanism of pain. How do pain killers work?

Teaching Suggestions

- Students find discussion of neurotransmitters fascinating if you throw in "real-life" examples of drugs, painkillers, and other chemicals which affect the synapse. For example, if the neurotransmitter receptors are blocked because of competing elements, no nerve impulse occurs in the postsynaptic cell. Examples: neurotoxic snake venom, curare (phytotoxin from bark of plants). Inhibition of the enzyme which breaks down acetylcholine causes synaptic problems, seen in the effects of cyanide, insecticides (malathion, parathion), and nerve gas. If the release of acetylcholine is inhibited, it can't leave presynaptic cells. An example is the toxin produced by *Clostridium botulina*, the toxin involved in botulism.

Stimulants like caffeine facilitate synaptic transmission in the CNS, as they block inhibitors. Amphetamines stimulate release of excitatory transmitters. On the other hand, tranquilizers deplete stores of norepinephrine in the CNS.

Suggested Readings

Selkoe, D.J. "Amyloid protein and Alzheimer's disease." *Scientific American*. November 1991. 68-78.

Various authors. Special issue of *Scientific American*; "The mind and the brain." September 1992. A special issue with articles covering the gender differences in the brain, neural disorders and the functioning of the brain.

Chapter 34. Responsiveness: Nervous Systems

Chapter Overview

The nervous system of invertebrates reflects their symmetry; radial animals have radial nerve nets, bilateral animals are characterized by increasing cephalization. Segmented animals may have a segmented nervous systems. The most highly evolved nervous system in invertebrates is seen in the octopus.

Vertebrates are characterized by a well developed central and peripheral nervous systems. The central nervous system consists of the brain and spinal cord. The peripheral nervous system may be subdivided into somatic (voluntary) and autonomic (involuntary) portions. The autonomic nervous division has both sympathetic nerves (typically excitatory), and parasympathetic nerves (typically inhibitory).

The hindbrain of the vertebrate brain is composed of the medulla, pons and cerebellum, and contains vital centers involved with respiration and vasomotor functions, and control of skeletal muscles. The midbrain is particularly important in processing visual information in fish and amphibians. In birds and mammals, the forebrain is the prominent control center; composed of the thalamus; the sensory relay center, the hypothalamus; the autonomic control center, and the cerebrum. The spinal cord, with both ascending and descending tracts, connects the brain and the peripheral nerves.

The human cerebrum is divided into two hemispheres, which have white matter inside, and the gray matter of the cortex outside. Complex foldings of the cortex increase the surface area of the cortex, which is the primary integrative area. The cortex has been mapped; visual, motor, and association areas are well known. The corpus callosum connects the two hemispheres. The RAS is involved in maintaining consciousness and allowing sleep. The limbic system is referred to as the "emotional brain." Learning involves many structures of the brain, and involves attention, short-term and long-term memory. Memory may be the result of reverberating circuits.

The peripheral nervous system includes the somatic system (cranial and spinal nerves) and the autonomic system, which is involved in maintaining homeostasis.

Lecture Outline

<u>Invertebrate nervous systems</u>

- Sponges have no nerve cells
- Cnidarians have nerve nets, neurons are not well organized
  - Impulses may be transmitted in both directions
- Echinoderms also have radially arranged nervous systems
- Bilaterally symmetrical animals have bilateral nervous systems

Trends in evolution:
- Increased number of nerve cells
- Nerve cells form the ganglia and brain; axons are collected in nerves
- Specialization of nerve cells
- Increased number of synapses, association neurons
- Cephalization; sense organs concentrated at one end

Flatworms with a ladder-type nervous system, cerebral ganglia

Annelids and arthropods have segmented ganglia, paired ventral nerve cords

Molluscs have at least 3 pair of ganglia
- Octopods have highly evolved nervous systems

<u>Vertebrate nervous systems</u>

- Functionally divided into CNS and PNS
- CNS is composed of the brain connected to a single, dorsal, hollow nerve cord
- PNS is composed of sensory structures and nerves
  - Afferent nerves are sensory
  - Efferent nerves are motor
    - Somatic (voluntary) nerves
    - Autonomic (involuntary) nerves
      - Sympathetic (typically stimulatory) division
      - Parasympathetic (typically inhibitory) division

<u>The vertebrate brain</u>

- Hindbrain
  - Medulla
    - Is continuous with the spinal cord
    - Contains vital centers; controls respiration, vasomotor functions
  - Cerebellum
    - Highly developed in birds and mammals
    - Involved in control of skeletal muscles
  - Pons
    - A bridge between the medulla and cerebellum and midbrain
    - Contains centers which regulate respiration
- Midbrain
  - Largest portion of brain in fish and amphibians
    - Optic lobes interpret visual information
    - In higher vertebrates, cerebrum interprets visual information
  - Other functions; responses to visual and auditory stimulation
- Forebrain
  - Thalamus
    - Relay center, sorts sensory messages
  - Hypothalamus
    - Control of the autonomic nervous system
    - Regulates the pituitary gland
    - Involved in thirst, hunger, pleasure, aggression

- Brainstem is composed of medulla, pons, midbrain, thalamus and hypothalamus
- Olfactory bulbs
    - Enlarged in many vertebrates
- Cerebrum
    - Two hemispheres
    - Composed of inner white matter (axons)
    - Outer layer is the cerebral cortex (gray matter; cell bodies and dendrites)
    - In mammals, cerebral cortex is highly folded, makes up most of brain mass
        - Folding is correlated with intelligence

<u>The human brain</u>

- Brain is covered by three layers of connective tissue; the meninges
    - Cerebrospinal fluid buoys the brain, cushions it, and nourishes brain tissue
- Spinal cord
    - Hollow, encased in vertebral column
    - Functions:
        - Control reflexes
        - Transmit messages up to brain (ascending tracts)
        - Carry messages from the brain (descending tracts)
        - Many neurons cross sides in the spinal cord
    - Outer white matter, inner H-shaped gray matter
        - Gray matter is composed of cell bodies and dendrites
        - White matter is myelinated axons
- Cerebrum
    - Weighs about 3 pounds, 25 billion neurons, 10^{14} synapses
    - Brain receives 20% of the blood; constant requirement for oxygen
        - Disruption of blood flow is a stroke (CVA)
    - Divided into 2 hemispheres, each into 5 lobes
        - Sensory areas in parietal lobes
        - Visual areas in occipital lobes
        - Motor areas in frontal lobes
            - Cortical area devoted to particular areas proportional to complexity of movement, or sensory sensitivity
    - Corpus callosum connects two hemispheres (white matter)
- RAS (reticular activating system) consists of neurons in the brainstem and thalamus; maintains consciousness
    - Damage to RAS leads to a coma
    - Slowing of signals leads to sleep
        - REM sleep (rapid eye movement) associated with dreams
        - Non-REM is normal sleep
        - Sleep also influenced by hypothalamus
- Limbic system surrounds the brain stem
    - Involved in emotions

EEG (electroencephalogram) measures brain action potentials
- Alpha waves occur at 10/sec; activity from visual area when at rest
- Beta waves result from mental activity
- Delta waves are associated with dreams

Learning has been demonstrated in animals in every phylum
- Sensory memory
  - Attention; information is lost in about 1 second
- Short-term memory; may be based on reverberating circuits
- Long-term memory
  - Has an unlimited capacity
  - Involves many brain structures
    - Hippocampus (part of limbic system) and thalamus
    - General interpretive area helps to put it all together
      - Wernicke's area- language interpretation

<u>Peripheral nervous system</u>

Somatic system
- Receptors which respond to environmental changes; sensory neurons and motor neurons which run to skeletal muscles
- 12 pairs of cranial nerves in mammals
  - Largest nerve, the vagus transmits information from many internal organs
- Humans have 31 pairs of spinal nerves
- Each spinal nerve has a dorsal (sensory or afferent) root, with a dorsal root ganglion (site of cell bodies of sensory neurons)
- Motor neurons exit via the ventral or efferent root
  - Cell bodies of motor neurons are located in the gray matter of the spinal cord

Autonomic system
- Maintains homeostasis
- Efferent portion is subdivided:
  - Sympathetic nerves are typically excitatory
  - Parasympathetic nerves are typically inhibitory
  - Act in opposite ways

Drugs may change neurotransmitters
- Continued use results in tolerance, and may finally cause addiction

Research and Discussion Topics

• Draw the brains of a shark, fish, frog, bird, cat and human. Analyze the relative sizes of the various parts of the brain and relate those to the differences between the animals.

• Describe a disorder of the brain, and relate the anatomy and physiology of the brain to the disease symptoms and treatment for the disorder. Possible topics: Alzheimer's or Parkinson's disease, cerebral palsy, multiple sclerosis.

Teaching Suggestions

- Students are often fascinated by the diagrams showing the areas of the body innervated by the motor and sensory cortices. Students may get a chuckle when they notice the close proximity of the sensory areas of the genitals and the toes- this may explain the sexual connection with the feet!

Suggested Readings

Nash, J.M. "Sizing up the sexes." *Time*. January 20, 1992. 42-51. Differences in the brains of males and females.

Chapter 35. Sensory Reception

Chapter Overview

Sensory receptors respond to varying types of stimuli, and ultimately transmit an action potential to the central nervous system. Receptors may be classified as mechanoreceptors, chemoreceptors, photoreceptors, thermoreceptors and electroreceptors, depending on the type of energy they respond to. Further, they may be classified by the source of their stimulant; as exteroceptors, proprioceptors and interoceptors.

Receptors are stimulated and produce a graded receptor potential which may result in an action potential. The action potentials in all sensory neurons are identical. The sensation is accomplished in the brain. Continued stimulation may result in sensory adaptation.

Mechanoreceptors include lateral line organs, and statocysts. Other mechanoreceptors are located in the skin (touch receptors), in muscles and joints (proprioceptors) and in the saccule, utricule and the semicircular canals of the inner ear (organs of equilibrium). Perhaps the most complicated mechanoreceptors are those of the cochlea. Sound energy which reaches the ear is passed to the inner ear, and vibrations of the fluid result in stimulation of hair cells, which is the stimulus of "hearing."

Chemoreceptors include the cells of taste buds and olfactory epithelium, and are highly developed in mammals. Specialized thermoreceptors are important for both ectoparasitic arthropods and some snakes. All vertebrates have thermoreceptors in their skin and some internal organs, including the hypothalamus. Electroreceptors are found in sharks and relatives, and some bony fish, which allow them to find prey more efficiently.

Invertebrate photoreceptors include ocelli, simple eyes, compound eyes seen in arthropods, and the camera-type eye seen in a few molluscs. The camera-type eye which evolved in octopods and vertebrates exhibits convergent evolution.

Vertebrate eyes have photoreceptors located in the retina, and a lens to focus light on the retina. In higher vertebrates, accommodation is accomplished by changing the shape of the lens. Light passes through the cornea, the aqueous humor, the lens, the vitreous body, then stimulates the rods and cones of the retina. Neurons of the eye pass visual impulses via the optic nerves to the visual cortex of the brain.

Lecture Outline

Sensory receptors

- Mammals have 5 classic senses: sight, hearing, smelling, taste and touch
  - Also should include balance
  - Touch is actually detection of pain, pressure and temperature
- Can classify sensory receptors by what they sense
  - Mechanoreceptors respond to mechanical energy
  - Chemoreceptors respond to various chemicals
  - Photoreceptors respond to light energy
  - Thermoreceptors respond to heat
  - Electroreceptors respond to electrical energy
- Can also classify them by the location of the stimuli to which they respond
  - Exteroceptors respond to external stimuli
  - Proprioceptors respond to stretching within muscles, tendons and joints
  - Interoceptors respond to other internal stimuli
- Sensory receptors transduce the energy that stimulates them into a receptor potential
  - It is a graded response, and fades as it passes down the dendrite
  - The receptor potential may generate an action potential
  - A strong stimulus may result in more action potentials
- Sensation takes place in the brain
  - Action potentials from the receptors are interpreted
  - Not all receptors cause conscious sensations (e.g. some chemoreceptors)
- Sensory receptors adapt to stimuli
  - Repeated stimuli result in decreased frequency of action potentials
  - Some adapt quickly, like olfactory receptors
  - Others, like cold or pain receptors adapt slowly, or not at all

Mechanoreceptors

- Lateral line organs are found in fish and respond to changes in water pressure or currents
- Statocysts are gravity receptors found in many invertebrates
- Touch receptors
  - Bare dendrites in the skin
  - Merkel's discs and Meissner's corpuscles are stimulated by light touch
  - Pacinian corpuscles are stimulated by deep pressure
  - Invertebrates with exoskeletons have sensory bristles
  - Vertebrates also have touch receptors at the base of hairs
- Proprioceptors allow positional sense
  - Stretching in muscles and joints
    - Muscle spindles detect muscle movement
    - Golgi tendon organs detect tendon movements
    - Joint receptors detect ligament movements
  - Important in maintaining balance
  - Very active receptors; not conscious of their sensations

- Labyrinth of the ear
    - Found in the inner ear
    - The saccule and utricle have gravity detectors
        - Otoliths- crystals of calcium carbonate
        - Receptors are hairs cells covered by the cupula
        - Stimulated when head changes position or by linear acceleration
    - Semicircular canals
        - In sagittal, coronal and transverse planes
        - Filled with endolymph
        - Ampulla has hair cells in a clump, the crista
            - Lack otoliths; respond to fluid movement
            - Respond to movement of head
        - Cause motion sickness when stimulated in an unfamiliar way
- Cochlea of the ear
    - Hearing is well developed in birds and mammals
    - Sound waves hit the tympanic membrane (eardrum)
    - Vibrations transferred to the hammer, anvil and stirrup
        - Middle ear filled with air
    - Stirrup transfers vibration to inner ear
    - Cochlea
        - Hair cells detect pressure waves
        - 3 canals separated by membranes, all filled with fluid
        - Top canal- vestibular canal is in contact with the oval membrane
            - The stirrup (stapes) presses against the oval membrane
        - Vestibular canal is continuous with the bottom canal, the tympanic canal
        - Middle canal houses the hair cells; the organ of Corti
            - Organ of Corti is overlain by the tectorial membrane; lies on top of the basilar membrane
        - Vibrations in the vestibular canal cause the basilar membrane to vibrate; hair cells are rubbed against tectorial membrane
        - Send impulses to brain

<u>Chemoreceptors</u>

- Mammals have taste buds; primarily located on the tongue
    - Taste buds are located at the bases of the papillae
    - Sense dissolved chemicals
    - Four basic tastes: sweet, sour, salty and bitter
- Olfactory epithelium is located in the upper nasal cavity
    - Sense odors in the air

<u>Thermoreceptors</u>

- Many ectoparasites use heat receptors to find endothermic prey
    - Mosquitoes, ticks
- Pit vipers (like rattlesnakes) and boas have thermoreceptors to find prey
- Vertebrates have cutaneous thermal receptors
    - Cold receptors are free nerve endings

Heat receptors are thought to be encapsulated dendrites
The hypothalamus also detects temperature

<u>Electroreceptors</u>

Found in sharks and some predatory bony fish
- Detect prey from electrical fields

Electric eels and rays can stun prey with electric organs

<u>Photoreceptors</u>

Rhodopsins are pigments which are photosensitive

Invertebrate photoreceptors
- Ocelli are able to detect light but not form images
  - Seen in cnidarians and flatworms
- True eyes have lenses which focus light on photoreceptors
  - Also need a brain for interpretation
- Compound eyes of arthropods
  - Eye surface is faceted, because of many ommatidia
  - Each produces a complete image
  - Brain interprets the mosaic of images
- Camera eye of squid and relatives, and vertebrates
  - Are analogous structures

Vertebrate eyes
- Structures:
  - Iris regulates light entering the eye through the pupil
  - Retina is the photoreceptive layer
  - Choroid layer is pigmented layer
  - Sclera is outer protective layer
  - Cornea is clear outer layer in anterior
  - Lens focuses light
  - Chamber between cornea and lens is filled with aqueous humor
  - Chamber behind lens is filled with the vitreous body (humor)
- Accommodation
  - Changing the curvature of the lens by the ciliary muscle
  - Nearsightedness is caused by a long eyeball; concave lenses correct
  - Farsightedness is caused by a short eyeball; convex lenses correct
- Vision
  - Retina contains rods (detect light) and cones (detect colors)
  - Cones are concentrated in the fovea
  - No photoreceptors are found in the area of the optic nerve = "Blind spot"
  - Light stimulates photoreceptors
    - Rhodopsin in rod cells, similar chemical in cones changes from cis-retinal to trans-retinal, then into components; opsin and retinal
    - Initiates an action potential; transmitted to the brain

Optic nerves exit eyes; meet and exchange fibers at the optic chiasma--> visual cortex of cerebrum

Research and Discussion Topics

- How does radial keratotomy work to correct nearsightedness? Research a newer method, known as orthokeratotomy.

- Discuss the causes and treatments of various eye diseases, such as glaucoma and cataracts.

Suggested Readings

Various authors. Special issue of *Scientific American*; "The mystery of sense." June 1993. Covers senses of touch, hearing, vision, smell and taste.

Koretz, J.F and G.H. Handelman. "How the human eye focuses." *Scientific American*. July 1988. 92-99. Aging and the human eye; accomodation and time.

Long, M.E. "The sense of sight." *National Geographic*. November 1992. 3-41. An excellent description of sight, disorders of the eye, and optical illusions.

Chapter 36. Internal Transport

Chapter Overview

Invertebrates have varying circulatory patterns; gastrovascular cavities, such as seen in cnidarians, and open circulatory systems, such as seen in arthropods and most molluscs. Insects have a more efficient system by the addition of trachea for respiration; annelids, cephalopods and echinoderms have closed circulatory systems, which also increase efficiency. A closed circulatory system with a pumping heart makes the vertebrate circulatory system also more efficient.

Vertebrates have blood which contains oxygen-carrying red blood cells, white blood cells which are phagocytic or produce antibodies or other substances involved in immunity, and platelets which function in clotting.

Blood vessels include arteries which carry blood from the heart, which diverge to form capillaries, the site of diffusion of materials between the blood and tissues. Veins ultimately return blood to the heart. The heart itself is the double pump which pumps blood to the pulmonary and the system circuits.

The heart beat is initiated by specialized cells in the right atrium, the SA node. The electrical impulse spreads over the atria and signals the AV node to send the impulse down to the ventricles, stimulating ventricular systole. The rhythmic beating of the heart and the force of contraction is regulated by the medulla and various hormones. The force of ventricular systole results in blood pressure, which is measured in an artery in the arm, and expressed as systolic pressure over diastolic pressure, in mm Hg.

In mammals, the four chambered heart pumps blood both to a pulmonary and systemic circuit. The coronary circuit, which supplies the heart muscle is a branch of the systemic circuit. The pulmonary trunk carries blood from the right ventricle to the pulmonary arteries which deliver blood to the lungs for oxygenation. The blood returns to the heart via the pulmonary veins. Blood from the left ventricle flows into the aorta, which has branches which supply the upper and lower body. Ultimately blood returns to the heart via the superior and inferior vena cava, which deliver deoxygenated blood to the right atrium.

The lymphatic system provides a way for tissue fluid to be returned to the venous system, as well as being a site of immune surveillance. Lymph capillaries pick up interstitial fluid and transport the lymph through one-way vessels, ultimately emptying into the subclavian veins. Lymph nodes are located on the lymph vessels, and lymphocytes in the lymph nodes aid in immunity and fighting infection.

Lecture Outline

Internal transport mechanisms

- Most have a circulatory system
  - Components:
    - Blood; a fluid connective tissue
    - A pump (a heart)
    - Vessels or spaces through which blood circulates
- Some invertebrates have no circulatory system
  - Depend on diffusion for exchange of materials
- Some combine digestive and transport functions
  - Cnidarians have a gastrovascular cavity
- Open circulatory systems
  - Seen in arthropods and most molluscs
  - Heart pumps blood into open-ended vessels
    - Blood flows into the hemocoel
    - Not as efficient as a closed system
    - Insects separate circulation of gases from the circulation of blood by the addition of a tracheal system
- Closed circulatory systems in invertebrates
  - Seen in annelids, cephalopods, echinoderms
  - Blood always in vessels
  - Small vessels allow diffusion of materials
  - May lack a heart (annelids), but have contractile vessel walls
  - Some have hemoglobin

Vertebrate circulatory systems

- All have a ventral muscular heart
- All have a closed circulatory system composed of:
  - A heart
  - Blood
  - Blood vessels
  - Lymph nodes, vessels and organs
- Functions:
  - Transport nutrients and wastes
  - Transport dissolved gases
  - Transport hormones
  - Maintain osmotic balance
  - Defense against microorganisms
  - Thermoregulation in endotherms

Blood

- Plasma is the fluid component
  - 92% water
  - Plasma proteins
    - Albumins which maintain osmotic balance of blood, others are transport proteins

        - Globulins include antibodies
        - Fibrinogen and other proteins are involved in clotting
            - Serum is plasma without the clotting proteins
- Red blood cells
    - Erythrocytes; the most abundant cellular components
    - Hemoglobin functions in transporting oxygen
    - Also carries carbon dioxide
    - Formed in red bone marrow
    - Short life span; about 120 days
        - Old RBC's are destroyed through the liver and spleen
    - Deficiency of hemoglobin causes anemia
        - Causes: hemorrhage, decreased production of either hemoglobin or red blood cells, or hemolytic anemias
- White blood cells
    - Leukocytes defend the body against foreign materials
    - Agranular leukocytes
        - Granules are not visible with the light microscope
        - Lymphocytes are involved in the immune response
        - Monocytes may become the macrophages in connective tissue
    - Granular leukocytes
        - Obvious granules in the cytoplasm
        - Neutrophils are phagocytes
        - Eosinophils are involved in allergic reactions
        - Basophils release histamines and heparin
    - Differential count looks at the numbers of different leukocytes; may give clues as to the cause of infection
    - Leukemia is a cancer of the white blood cells
        - WBC's are immature, and nonfunctional
        - Anemia is also a problem
- Platelets
    - Cell fragments, formed in the bone marrow
    - Important in hemostasis (clotting)
    - Platelets adhere to the cut edges of a blood vessel
    - Various chemicals interact to form the clot
    - Hemophiliacs lack one of the clotting factors
    - Clotting also requires vitamin K, calcium, fibrinogen and prothrombin

<u>Circulatory vessels</u>

- Arteries
    - Carry blood away from the heart
    - Typically carry oxygenated blood
    - When they enter organs, they branch into arterioles, then capillaries
    - Arteriole walls are muscular; can vasoconstrict or vasodilate
- Capillaries
    - Very small vessels- only slightly larger than the diameter of a RBC
    - Walls are one cell thick
    - Site of diffusion of materials

- Veins
  - Carry blood back to the heart

<u>The heart</u>

- 4 chambered in mammals
  - Two atria, two ventricles
- Covered with 2 layers of pericardium; pericardial fluid between layers
- Valves prevent backflow of blood
  - Atrioventricular valves are the mitral and tricuspid valves
  - Semilunar valves are between ventricles and great arteries
- Heartbeat is initiated by the sinoatrial node (pacemaker)
  - Electrical impulses travel to the atrioventricular valve
  - Impulse causes ventricles to contract
- Cardiac cycle
  - Ventricular contraction is systole
  - Ventricular relaxation is diastole
  - Pulse is a reflection of ventricular systole
  - Factors which affect cardiac cycle:
    - Sympathetic stimulation causes increase in strength of heart beat
    - Epinephrine and norepinephrine also increase heart beat
    - Athletic training increases size of heart, and also cardiac output
  - Heart sounds
    - Lub sound marks closure of AV valves, beginning of ventricular systole
    - Dup sound marks closure of semilunar valves, beginning of ventricular diastole
    - Heart murmurs may be due to valve dysfunction
- Blood pressure
  - Due to resistance in vessels of blood flow
  - Higher blood pressure may be due to increased blood volume, heredity obesity, salt intake
  - Expressed as systolic over diastolic
  - Normal pressure is 120/80 mm Hg
  - Pressure is highest in arteries
  - Pressure is low in capillaries and veins
    - Veins have valves to prevent backflow
    - Varicose veins are flabby veins
    - Hemorrhoids are varicose veins in the rectum
  - Baroreceptors in arteries sense blood pressure; send nervous messages to medulla
  - Angiotensins are hormones involved in vasoconstriction
    - Produced in response to renal hormone renin

Circulatory circuits

- Pulmonary circulation
  - Blood from the right ventricle passes to the pulmonary trunk--> pulmonary arteries--> lungs
  - Function: pick up oxygen, drop off carbon dioxide
- Systemic circulation
  - Blood from the left ventricle passes to the aorta
  - Branches of the aorta:
    - Carotid arteries supply the brain, subclavian arteries to the arm and shoulder region
    - Coronary arteries supply the heart muscle itself
    - Mesenteric arteries supply the intestines, renal arteries supply the kidneys, iliac arteries supply the legs
  - Systemic veins:
    - Jugular veins carry blood from the brain, subclavian veins carry blood from the arms and shoulders
      - Drain into the superior vena cava--> right atrium
    - Renal veins drain the kidneys, hepatic veins from the liver, iliac veins drain the legs; all drain into the inferior vena cava--> right atrium
  - Coronary circulation
    - Coronary arteries branch off of the aorta; supply the heart muscle
    - Coronary veins form the coronary sinus which drains into the right atrium
    - Blockage of arteries causes heart disease
  - The hepatic portal system processes blood from the intestines
    - Intestinal veins merge to form the hepatic portal vein
    - Capillaries within the liver process blood
      - Store nutrients, detoxify toxins

Lymphatic system

- Functions:
  - Collect and return tissue fluid to the venous system
  - Defend against foreign cells, viruses
  - Absorb fats from the intestines
- Lymph nodes (lymph glands)
  - Functions: filter lymph, lymphocytes reside there
- Lymphatic organs
  - Tonsils are lymph tissue in the back of the oral cavity and throat
    - "Adenoids" are tonsils in the nasopharynx
- Lymph vessels
  - Small lymph capillaries pick up interstitial fluid (tissue fluid)
    - Tissue fluid is forced out of circulatory capillaries by osmotic pressure of the blood
  - Lymph capillaries pick up about 10% of the interstitial fluid and proteins
  - Ends of lymph capillaries have tiny valves; let fluid in

Blockage of lymphatic vessels leads to edema
Lymph capillaries merge to form the larger thoracic and right lymph ducts which drain into the subclavian veins

Research and Discussion Topics

- Describe the evolution of the heart from fish to amphibians and reptiles to the four-chambered heart seen in birds and mammals.

- Discuss the three different types of arteriosclerosis, including atherosclerosis, which is the most common form. What are the proposed causes of arteriosclerosis?

Teaching Suggestions

- I'm always amazed at the lack of knowledge about the heart! Students don't typically know the circuits, and the term heart attack is the only other thing they know about circulation. Plan to spend time diagraming the heart and the circuits. Be certain that students understand the importance of the coronary circuit.

Suggested Readings

Pines, M. Editor. Blood: Bearer of life and death. A report from the Howard Hughes Medical Institute. Available (free!) from the Howard Hugues Medical Institute, 6701 Rockledge Dr, Bethesda, MD 20817. 59 pages. A complete description of blood, sickle cell disease, hemophilia, blood clotting.

Krupka, L.R. and A.M. Vener. "College students' knowledge of cardiovascular disease: implications for the biology teacher." *American Biology Teacher*. October 1991. 53 (7) 394-398. Points out gaps in knowledge of cardiovascular anatomy, physiology and health issues in college students.

Zivan, J.A. and D.W. Choi. "Stroke therapy." *Scientific American*. July 1991. 56-63. Current treatments for strokes, including tPA.

Chapter 37. Internal Defense: Immunity

Chapter Overview

All animals have the ability to recognize between self and nonself, and defense mechanisms may be classified as nonspecific and specific defense mechanisms. Invertebrates typically exhibit simple nonspecific responses, such as phagocytosis and the inflammatory response. Vertebrates have nonspecific defenses as well, including phagocytic cells, production of interferons, and the inflammatory response. The specialized lymphatic system which houses specialized white blood cells allows for a complex specific defense against pathogens.

Involved in antibody-mediated immunity, T cells and B cells have complex functions and responses to pathogens such as bacteria, viruses and fungi. In short, macrophages which phagocytize pathogens display the antigen of the pathogen on their cell membranes, along with their own MHC molecules. These interact with both T cells and B cells to activate them. B cells divide rapidly to become plasma cells, which produce antibodies specific to that presented antigen. Antibodies are complex molecules, with constant regions, and variable regions, which bind to antigens like a lock and key.

Antibody-antigen complexes stimulate further interactions between pathogens and leukocytes, including the activation of complement, which is a non-specific response.

Cell-mediated immunity involves T cells and macrophages, which also have very complex modes of presentation and activation to destroy viruses and cancer cells. T cells differentiate into cytotoxic, helper, suppressor and memory T cells, which have various functions.

The primary response to a pathogen involves the antibody-mediated and cell-mediated immune responses. The presence of memory cells allows the secondary immune response to be mounted quickly, before pathogens cause noticeable disease. This is active immunity. Active immunity may occur naturally, or by a vaccination. Passive immunity is relatively short-lived and involves the acquisition of antibodies via a gamma globulin shot or via the placenta or breast-feeding.

Cancer cells may be targeted by macrophages, cytotoxic T cells and natural killer cells. Current research involves the use of monoclonal antibodies which specifically target the cancer cells.

Immunological response to the MHC antigens of transplanted tissue may cause rejection. Autoimmune diseases involve an inappropriate immune response to one's own tissues, such as in multiple sclerosis and rheumatoid arthritis. Allergic reactions are immune responses to common non-pathogenic allergens, such as dust or pollen. After sensitization, mast cells cause inflammation, and in a severe reaction, may cause death by widespread vasodilation.

AIDS is a disease caused by HIV, which specifically targets helper T cells. By destroying a component of the antibody-mediated immunity, persons with HIV are particularly susceptible to other pathogens. AIDS is particularly hard to treat, as it crosses the blood-brain barrier, mutates rapidly, and does not infect other animals which would allow laboratory testing.

Lecture Outline

Defense mechanisms

- Pathogens are organisms that cause disease
  - Recognized as "nonself"
  - Antigens stimulate an immune response
    - Includes proteins, RNA, DNA, and some carbohydrates
- Nonspecific defense mechanisms
  - Phagocytosis of invading bacteria
- Specific defense mechanisms
  - Includes immune response; are specific to each antigen
  - Antibodies cause ultimate destruction of foreign cells

Invertebrate defense mechanisms

- Typically nonspecific defenses, such as phagocytosis and inflammatory response
- Phagocytes in the coelom
- Some annelids and cnidarians have some specific immune mechanisms and immunological memory
- Echinoderms have differentiated white blood cells

Vertebrate nonspecific defense mechanisms

- Epithelia are barriers to pathogens
  - Lysozymes attack bacteria
- Acids and enzymes of the stomach destroy ingested pathogens
- Phagocytes in the respiratory pathways
- Interferons are a defense against viruses
  - Signal other cells to prevent viral replication
  - May also stimulate macrophages
  - Recombinant techniques produce interferons
- Inflammation aids in defense
  - Injured cells and basophils release histamines; blood vessels dilate
    - Blood flow to infected area increases; becomes red and warm
  - Blood vessels become more permeable; white blood cells leave the capillaries
  - Edema causes pain
  - Inflammation is local; fever is widespread inflammatory response
    - Interleukin-1 (IL-1) resets hypothalamic thermostat

Fever decreases growth of microorganisms
Also increases production of T cells and antibodies
Phagocytes
Neutrophils and macrophages phagocytize bacteria

Specific defense mechanisms

Nonspecific defense mechanisms operate first
Specific defense mechanisms require several days
Two types:
Antibody mediated immunity; lymphocytes produce antibodies
Cell-mediated immunity; lymphocytes attack pathogens
Lymphocytes
T cells
Originate in bone marrow
Become immunocompetent in the thymus
Cytotoxic T cells (killer T cells) recognize and destroy pathogens
Helper T cells activate the immune response
Suppressor T cells release cytokines
B cells
Produced in the bone marrow, also mature there
Specialized to respond to a different antigen
When contacts target antigen, divides forming a clone of plasma cells
Plasma cells function in antibody production
Macrophages
Ingests pathogens, some of the bacterial antigens are displayed on the surface of the macrophage
Referred to as an APC (antigen presenting cell)
This activates helper T cells
Also produce interferons and interleukin-1
IL-1 activates B and helper T cells
The thymus
T cells become immunocompetent
Produces thymosin, believed to stimulate T cells
MHC; major histocompatibility complex
Allows identification of self
Human MHC is the HLA group (human leukocyte antigen)
More than 100 genes; everyone has a different "blueprint"

Functioning of antibody-mediated immunity (humoral immunity)

B cells produce antibodies
Competent B cells match their receptors with the antigen of the pathogen
Recognize the antigens of the pathogen presented by the macrophage
Form a complex with the MHC of the macrophage and the pathogenic antigens

- T cells are activated by recognizing the antigen/MHC complex on the macrophage, also stimulated by the IL-1 from the macrophage
    - T cells interact with the B cells
- B cell is now activated
    - Increases in size, divides by mitosis--> clone of identical cells
    - Clonal selection
        - Some mature into plasma cells, which secrete the specific antibody for that pathogen
        - Others become memory cells
            - Continue to produce small amounts of that antibody
            - All of the antibodies in plasma are the gamma globulins
            - New infection of same pathogen stimulates memory cells to divide and form new clones of plasma cells

Antibodies

- Also known as immunoglobulins (Ig)
- Are proteins produced in response to antigens
    - Term antigen means "antibody-generator"
- Function: to bind to the antigen
- Antigenic determinant
    - The shape which is recognized by the antibody of T-cell receptor
    - May have 5- over 200
- Haptenes are substances which are not antigenic but do stimulate an immune response
- Shapes of antibodies
    - Y shaped
    - Two arms are binding sites
        - One antibody can bind with two antigen molecules
    - Tail of antibody functions in binding to cells or activating complement
    - Polypeptide chains
        - Two heavy chains
            - Are long chains; are identical
        - Two light chains
            - Are short chains; are also identical
        - Each chain has a constant segment, the C region
            - Amino acid sequence is the same
            - In B cell receptors, this anchors the molecule to the cell
        - The junctional segment, the J region
            - Variable amino acid sequences
        - The variable segment, the V region
            - Complex 3-D structure
            - This is the unique section of the antibody
            - This is the specific part that matches the antigen
                - V region and antigen fit like lock and key
            - In B cell receptors, this extends from the cell

Antibody classes

IgG

The primary immunoglobulins in humans
Composes 75% of the gamma globulin fraction of the plasma
Defense against bacteria, viruses, and some fungi
Stimulates macrophages and activate complement

IgM

Defends against bacteria, viruses, and some fungi, with IgG
Stimulates macrophages and activate complement, with IgG

IgA

Present in mucus, tears and saliva, defense against ingested or inhaled pathogens
Prevents viruses and bacteria from attaching to epithelia

IgD

Present on B cells, aids in antigenic binding

IgE

Stimulates the release of histamines

Binding of antibodies to antigens

Form a clump of antigen-antibody complexes
This may inactivate the pathogen or its toxin
Stimulates phagocytes to ingest pathogen
Works through the complement system

Complement

20+ proteins in plasma, other body fluids
Normally inactivated
IgG and IgM antibodies fix complement (make them functional)
May digest pathogenic cell
May coat the pathogen, aiding in the phagocytic process
Also increases inflammation
Are not specific; act on any antigen-antibody complex

<u>Cell-mediated immunity</u>

T cells and macrophages destroy cells infected with viruses
T cells become activated when recognize the antigen/MHC complexes on cells infected with viruses
T cells increase in size, divide to form a clone
Cytotoxic T cells destroy target cells
Binds with antigen, secretes granules which destroy the target cell
May produce lymphotoxins which are toxic to cancer cells
Helper T cells and macrophages produce interleukins and interferons
Suppressor T cells <u>inhibit</u> activity of T cells, B cells and macrophages
Develop much more slowly
Responsible for the "end" of the immune response
Memory T cells act like memory B cells

Primary and secondary immune responses

- First exposure to an antigen causes the primary response
    - After exposure, lymphocytes build up antibodies, primarily IgM
- Secondary response involves memory cells
    - After second exposure, latent period is shorter
        - Predominant antibody is IgG
- Secondary response prevents pathogen from causing disease
- Booster shots of vaccines cause a secondary response

Active and passive immunity

- Active immunity is due to the activity of your own immune system
    - May be naturally induced, from getting an infection
    - May be artificially induced, by vaccinations
        - Vaccines are caused by weakened or killed pathogens, or altered toxins (which act as antigens)
- Passive immunity is the acquisition of antibodies
    - May be a serum or gamma globulin fraction injection
    - May be acquired from the mother while in the uterus, or via breast feeding
    - Effects are not long lasting; act for several months

Defense against cancer

- Cancerous cells are abnormal, in some ways present abnormal antigens
- NK cells and cytotoxic T cells defend against cancer cells
    - T cells produce interleukins, which attract and activate macrophages and NK cells
    - T cells also produce interferons, which suppress cancerous growth
- Macrophages produce TNF (tumor necrosis factor), which also suppresses cancerous growth
- Cancer cells may circumvent the immune system, as they are antigenically similar to normal cells
    - Blocking antibodies may block the T cells
- Monoclonal antibodies are B cells with antibodies to cancerous cells fused to cancerous cells
    - Can be cloned; are specific to the original cancer cells

Immunological rejection

- Grafts are rejected because the body recognizes the different MHC molecules
    - Cell typing attempts to match the MHC antigens
    - May use immunosuppressive drugs
    - Corneal transplants are successful because it has almost no capillaries or lymph vessels
- Autoimmune diseases are due to an inappropriate immune response
    - Examples: rheumatoid arthritis, MS, SLE, insulin dependent diabetes, psoriasis and scleroderma
    - Bacterial or viral infections often precede onset

Allergies are caused by the production of allergens
- Persons with allergies produce distinctive IgE immunoglobulins
- In sensitization, macrophages, T and B cells interact; plasma cells produce IgE in response to the allergen
- Mast cells are activated; they release histamines and seratonin, which cause blood vessels to dilate and leak
- Prolonged allergic response is caused by the migration of white blood cells to the inflamed area
  - They release chemicals which further the response
- Asthma affects the bronchioles of the lungs
- Other allergies affect the walls of the GI tract, cause diarrhea
- Hives is caused by an allergic reaction in the skin
- Systemic anaphylaxis occurs when a person develops a severe allergy to a compound, resulting in widespread vasodilation
- Treat allergies with antihistamines

AIDS

Caused by a retrovirus, HIV

Attaches to the CD4 protein on the helper T cell
- Over time, destroys the helper T cells
- Decreases ability of immune system to combat other pathogens

May pass the blood-brain barrier, and infect neurons

Treatment:
- AZT blocks reverse transcriptase

Virus mutates quickly; new antigens appear

Research and Discussion Topics

• Investigate a particular autoimmune disease. Describe the disease symptoms, treatments, and suggested causes. Subjects on which students should find sufficient research material: insulin dependent diabetes, multiple sclerosis, lupus.

• Discuss what is known on the activity of the complement system. How do these proteins aid in suppression of pathogens?

Teaching Suggestions

• Students find this subject intrinsically confusing. I find it helpful to first introduce "the players"; the various cells and their functions. Then I differentiate between cell-mediated and antibody-mediated immunity, and finally put it all together.

Suggested Readings

Special issue ot *Scientific American*. September 1993. Articles on grafting, cancer, AIDS, allergy, multiple sclerosis, rheumatoid arthritis etc.

Boon, R. "Teaching the immune system to fight cancer." *Scientific American*. March 1993. 82-89. Current research on tumor-rejection antigens.

Atkinson, M.A. and N. K. Maclaren. "What causes diabetes?" *Scientific American*. July 1990. 62-66. Description of the autoimmune response that results in insulin-dependent diabetes.

Caren, L.D. "Effects of exercise on the human immune system." *Bioscience*. June 1991. 41 (6): 410-414

Jaret, P. "Our immune system: the wars within." *National Geographic*. June 1986. 702-734. Very richly illustrated depiction of the immune system.

Rennie, J. "The body against itself." *Scientific American*. December 1990. 106-115. A description of autoimmune diseases.

Keister, E. "A little fever is good for you." *Science '84*. November 1984. 168-173. A description of fever in the immune response.

Renegar, K.B. and P.A. Small. "Monoclonal antibodies." *Carolina Tips* November 1986. A description of monoclonal antibodies, their production and use, and the kits that Carolina sells for student experiments.

Anderson, R.M. and R.M. May. "Understanding the AIDS pandemic." *Scientific American*. May 1992. 58-66. Comparing the mathematical models and the statistics about AIDS.

Diamond, J. 'The mysterious origin of AIDS. *Natural History*. September 1992. 24-29. The history of HIV, and the relation to SIV.

Chapter 38. Gas Exchange

Chapter Overview

All animals must have a source of oxygen; some obtain oxygen by diffusion through the body surface; most animals have specialized outfoldings (gills) or infoldings (lungs and spiracles) to maximize gas exchange. Animals may use flagella or cilia to move their medium over the respiratory structures, but most actively move entire parts of the body to move the water or air over the gill or lung. Small aquatic animals (or terrestrial animals with moist bodies) with a high surface-to-volume ratio may rely on cutaneous respiration. Trachea are unique, highly efficient tubes which directly pipe air to cells in insects and a few related arthropods.

Gills are very efficient structures seen in aquatic animals, ranging from dermal gills of echinoderms, the food-gathering gills of molluscs to the gills of fish, which have extensive surface areas.

Lungs are respiratory structures adapted for terrestrial life, and range from simple lungs seen in lungfish and amphibians, to complex structures with millions of alveoli seen in birds and mammals. Birds also have accessory air sacs which allow for a very efficient, flow-through respiratory system.

In humans, air enters the respiratory system through the nostrils, moves to the pharynx, past the epiglottis to the larynx, which functions in sound production, and then to the trachea, bronchi and bronchioles. The respiratory surface is the simple squamous lining of the terminal air sacs, the alveoli. The breathing process is controlled by centers in the medulla and pons, and inspiration is a result of contraction of the diaphragm and intercostal muscles. Levels of carbon dioxide indirectly influence the rate of breathing by changes in pH which are monitored by chemoreceptors in the medulla.

In the alveoli, due to the relatively high partial pressure of oxygen in the air sac, air diffuses through the epithelium to the blood. Hemoglobin has a high affinity to oxygen, and when bound, is referred to as oxyhemoglobin. Hemoglobin also carries a small amount of carbon dioxide, but most dissociates and forms bicarbonate ions, which are dissolved in the plasma.

Pollutants may be trapped in the cilia of the respiratory passageways, and are ultimately transported to the pharynx, where the particulates are swallowed. In the alveoli, pollutants may be phagocytized by macrophages. Ultimately, continued exposure to pollutants, including cigarette smoke, leads to chronic bronchitis and/or emphysema (COPD).

Lecture Outline

Respiratory structures

- Oxygen diffuses into an animal, and carbon dioxide diffuses out of the animal
  - May diffuse through the body surface
  - May have specialized structures
    - Aquatic animals often have gills
    - Terrestrial animals typically have trachea or lungs
  - Gases must be dissolved in water; lungs surfaces are moist
- Animals actively move air or water over or through the respiratory structure
  - Terrestrial animals lose water in this process
  - Sponges move water via flagellated cells
  - Most animals actively move water over their gills, or breathe air

Cutaneous respiration

- Seen in small animals with a high surface-to-volume ratio
  - Low metabolic rate
- Terrestrial animals must have a moist body surface
- Seen in molluscs, annelids and many amphibians
- Network of capillaries under the epidermis

Trachea

- Insects and a few other arthropods have tracheal tubes
- Openings are spiracles
- May actively pump air into trachea
  - Direct delivery of air to cells
  - Trachea are fluid filled at the ends

Gills

- Gills are evaginations of the body surface
- Are moist, delicate structures; buoyant in water
- Echinoderms have dermal gills with ciliated cells
  - Gases are exchanged between the gills and the coelomic fluid
- Most molluscs have gills
  - Bivalves have gills which also function in trapping food
- Chordate gills are typically pharyngeal
  - Bony fish have an operculum covering the gills
  - Gill filaments are the respiratory structures
  - Countercurrent flow maximizes respiratory efficiency
    - Flow of blood is opposite flow of water

Lungs

- Invaginations of the body surface or pharynx
- Spiders and horseshoe crabs have book lungs
  - Plate-like respiratory tissues
- Lungfish breathe via lungs
- Amphibians have simple lungs (but typically rely on cutaneous respiration)

Reptilian lungs are simple sacs
Bird lungs have air sacs as extensions of the respiratory system
Air flows through their respiratory system, not in and out
No dead air space
Countercurrent flow of air and blood
Mammalian lungs are characterized by an extensive surface area

<u>The mammalian respiratory system</u>

Air enters through the nostrils, flows through the nasal cavities
Air is filtered, warmed, moistened
Copious mucus production; propelled by cilia towards pharynx
Pharynx is the junction of the nasal and oral cavities
Site of passage of food as well as air
Epiglottis covers the opening to the respiratory passages during swallowing
Larynx is the "voice box"
Supported by cartilages, as is the trachea
Trachea is the "windpipe"
Cartilages are C-shaped; function in keeping the airway open
Trachea bifurcates into the primary bronchi
Cilia of trachea, bronchi beat upwards

<u>The mammalian lung</u>

Lungs are divided into lobes
Three right lobes, two left lobes (heart takes up space on that side)
Lungs are covered by pleural membrane, continuous with membrane lining the thoracic cavity
Pleural cavity filled with pleural fluid in between the two membranes
Bronchi branch within the lung--> bronchioles
Bronchioles lead to the alveoli
Alveoli are composed of simple squamous epithelium
Are surrounded by capillaries
Are the site of diffusion of gases

<u>Breathing and exchange of gases</u>

Inhaling is inspiration; exhaling is expiration
Inspiration caused by contraction of the diaphragm and chest muscles
Expiration is caused by relaxation of the diaphragm
Respiratory centers regulate respiration
Centers in the medulla and pons regulate breathing
Chemoreceptors respond to pH changes (high carbon dioxide levels cause a lowering of the pH)
Nerve impulses from the brain travel via the phrenic nerves (stimulate the diaphragm), and intercostal nerves (stimulate intercostal muscles)

Diffusion of gases in the alveolus
Gases diffuse based on their partial pressures
Higher partial pressure of oxygen in the inspired air
Oxygen diffuses from the air sacs to the blood
Higher partial pressure of carbon dioxide in the blood
Carbon dioxide diffuses from the blood to the air sacs
Oxygen is transported as oxyhemoglobin
One hemoglobin (Hb) molecule can transport 4 molecules of oxygen
Is a reversible reaction
Binding affected by pH, partial pressures of oxygen and carbon dioxide, and temperature
Oxygen-hemoglobin dissociation curves
As oxygen concentration increases, binding with Hb increases
Oxyhemoglobin dissociates in more acidic solutions
High carbon dioxide in tissues lowers the pH
Carbon dioxide is transported by hemoglobin, also dissolved in the plasma
20% is attached to hemoglobin
7% is dissolved in the plasma as CO_2
Remainder is in the form of bicarbonate ions
Reactions catalyzed by carbonic anhydrase

Altitude and pollutant affect respiration
Scuba divers may experience decompression sickness (bends) due to bubbles of gases forming in capillaries and joints
Pollution damages the ciliated cells which remove inhaled particulates
Smokers introduce additional particulates to the lungs
Macrophages engulf particulates
Chronic bronchitis and emphysema are clearly linked to smoking
Bronchitis is characterized by hyperproduction of mucus
Emphysema is a chronic degenerative disorder
Alveoli are destroyed; respiratory surface is decreased
Causes enlargement of the right ventricle

Research and Discussion Topics

• Fish evolved in freshwater habitats, and had lungs. Describe the development of the lung into the swim bladder, and the retention of the lung in the lungfish and lobefins (which shared an ancestor with the amphibians).

• Diving seals and whales have a variety of respiratory adaptations that allow them to dive deeply. Discuss these anatomic and physiologic adaptations.

• Discuss the problems seen in SCUBA divers, including nitrogen narcosis and decompression sickness. Why do human divers experience these potential problems, but diving whales and seals do not?

Teaching Suggestions

• This is a good point to deliver a "public service message" on the physiological effects of smoking.

Suggested Readings

Caldwell, M. "Resurrection of a killer." *Discover*. 1992. 59-64. Description of the multi-drug resistant strains of tuberculosis.

Rosenthal, E. "Return of consumption." *Discover*. June 1990. 80-83. Description of the disease, tuberculosis, and drug-resistant strains.

Schelling, T.C. "Addictive drugs: the cigarette experience." *Science*. 24 January 1992. 255: 430-133. A description of the physiological side of cigarette addiction.

Chapter 39. Processing Food and Nutrition

Chapter Overview

Simple animals digest food intracellularly, but most animals have a digestive cavity in which digestion occurs. Simple invertebrates have one opening to the digestive cavity, but most have a complete (mouth to anus) digestive tract, allowing more continuous feeding.

The human digestive tract has a series of specialized structures which aid in digestion and absorption of food. Food is chewed in the mouth with the aid of teeth and tongue, and salivary amylase begins digestion of polysaccharides. When a bolus of food is swallowed, it passes into the pharynx, past the epiglottis, into the esophagus, and peristalsis propels it toward the stomach. In the stomach, the bolus is transformed into chyme by mechanical churning, hydrochloric acid and some proteolytic enzymes. Little absorption occurs here.

The small intestine is the primary site of digestion and absorption, and most enzymes empty into the small intestine from the pancreas. Bile, a conglomerate fluid, is produced by the liver, is concentrated by the gall bladder, and ultimately also empties into the small intestine (duodenum). Villi and microvilli both increase the effective surface area of the small intestine. The large intestine is the site of final water reabsorption from the undigested material, and wastes are finally eliminated through the anus.

Nutrients are needed by organisms both for energy sources, and biosynthesis. Carbohydrates are the primary energy source, fueling aerobic respiration, and are ingested in the form of complex carbohydrates (starch can be digested, cellulose is the primary component of fiber) or simple sugars. Lipids may also be used as energy sources, and biosynthetically, as they are the major component of cell membranes. Saturated fats are primarily from animal sources, and are relatively unhealthy. Animal foods are also sources of cholesterol. Proteins are primarily used as building blocks to manufacture necessary proteins. Essential amino acids are those from which the body makes all 20 amino acids.

Vitamins are organic molecules, minerals are inorganic molecules; both of which are required by the body in small amounts. Sodium chloride is the primary salt in the body. When calories ingested exceed calories required, weight is gained. When calories burned exceed calories ingested, weight is lost. Malnutrition is defined as the lack of essential components in the diet. Worldwide, lack of protein and iron is a serious problem.

Lecture Outline

Obtaining nutrients

- All animals need food as an energy source; all are heterotrophs
- Animals are adapted in many ways for obtaining food
- Ingestion includes taking food in and swallowing it
- Molecules are then digested and absorbed
- Undigestible materials are egested or eliminated

Designs of digestive systems

- Simplest animals and some parasites have no digestive system at all
    - Digestion is intracellular within food vacuoles
- Simple invertebrates have one opening to the digestive cavity
    - Cnidarians have a gastrovascular cavity
    - Flatworms predigest food, then ingest it into a branched digestive cavity
- Most invertebrates and all vertebrates have a complete digestive tract

The human digestive system

- Hollow tube, from mouth to anus
- Wall has 4 layers:
    - Inner mucosa
    - Submucosa is richly vascularized
    - Muscular layer, typically one circular, one longitudinal
    - Adventitia, the outer connective tissue coat
        - After the esophagus, the adventitia is referred to as the visceral peritoneum
- Structures of the mouth
    - Teeth are specialized
        - Incisors for biting
        - Canines for tearing food
        - Premolars and molars for chewing food
        - Teeth are covered by enamel, dentin inside
    - Salivary glands secrete saliva into the mouth
        - Saliva moistens food, salivary amylase begins digestion of starch
- Pharynx and esophagus carry food to the stomach
    - Swallowing moves food bolus past epiglottis to esophagus
    - Bolus is moved by peristalsis in esophagus
- The stomach
    - Functions in food storage and initial digestion
    - Breaks down food to chyme
    - When unfilled, rugae (ridges) mark the inside
    - Glandular wall
        - Cells produce HCl
        - Pepsinogen (precursor of pepsin, a proteolytic enzyme)
        - Water, simple sugars, salts, alcohols some drugs are absorbed
    - Ultimately chyme is passed through the pylorus to the duodenum

- The small intestine
  - Primary site of digestion and absorption
  - Sections:
    - Duodenum; site of chemical digestion
      - Enzymes from pancreas, bile from liver empty here
    - Jejunum; absorption of nutrients
    - Ileum; further absorption of nutrients and water
  - Lined with villi
    - Cells of villi have numerous microvilli
    - Both act to increase the surface area of the small intestine
  - Nutrients are absorbed by the cells of the villi
    - Nutrients are passed to capillaries and lymph vessels
    - Most nutrient molecules are absorbed by active transport
      - Capillaries receive nutrient molecules; this blood is ultimately processed by the portal system in the liver
    - Lipids diffuse through intestinal cells and enter the lymph vessel
    - Lipids then directly enter the blood stream when the thoracic lymph vessel empties into the subclavian vein
- The liver
  - Located under the diaphragm
  - Complex actions
    - Secretes bile, which aids in emulsification of fats
      - Bile is a mixture of water, bile salts, pigments, cholesterol, salts and lecithin
    - Stores nutrients, some minerals and vitamins; converts glucose to glycogen, converts excess amino acids to fatty acids and urea
  - Detoxifies alcohol, drugs, toxins
- The gall bladder
  - Stores and concentrates bile, releases it when needed
- The pancreas
  - Secretes trypsins (break down polysaccharides to disaccharides), pancreatic lipase, amylase and nucleases
- The large intestine or colon
  - Absorbs water from the undigested material
  - The appendix is at the junction of small and large intestines
    - A blind, fingerlike projection
    - No function in humans, is a vestigial structure
    - Inflammation results in appendicitis
  - A cecum (pouch) is also at this junction
  - Parts of the large intestine: cecum, ascending, transverse, descending and sigmoid colon, rectum, anus
  - Bacteria; *E. coli* of the large intestine produce a few vitamins, which we absorb

Undigested material which passes from the large intestine is eliminated
Large intestine also functions in excretion of bile pigments
Malfunction of the large intestine leads to diarrhea or constipation
Cancer of the colon is linked to low fiber in the diet.

Nutrition

- Metabolism involves catabolism (breakdown) and anabolism (synthesis)
  - Calories are measurements of the amount of energy in food
  - In the body, nutrients are directly used by cells, or stored by the liver
- Carbohydrates
  - Sugars and starches are primary sources of energy
  - Not considered essential nutrients, because body can gain enough energy from fats and proteins (but <u>not</u> a healthy mix!)
  - Complex carbohydrates
    - Starches are digested into simple sugars
    - Cellulose is a primary component of dietary fiber
  - Excess monosaccharides are converted to glycogen by the liver
    - May also be converted fats and stored in that form
- Lipids
  - Necessary as part of the cell membrane, some hormones and bile salts
    - May also be used as an energy source
  - Three essential fatty acids are used to make all other lipids needed (including cholesterol)
    - Most fats in the diet are triglycerides
  - Triglycerides may be saturated (no double bonds)
    - Monounsaturated (one double bond)
    - Polyunsaturated (more than one double bond)
      - Monounsaturated fats may be the most healthy form; saturated fats are relatively unhealthy
      - Animal foods are high in cholesterol and saturated fats
      - Olive oil is a monounsaturated fat
- Proteins
  - Essential as building blocks, enzymes, components of molecules such as hemoglobin and myosin
  - Americans eat much more than the recommended daily intake
    - Most protein sources are high in fat
  - Essential amino acids are those which must be supplied in the diet
    - Complete proteins come from animal sources; have the appropriate mix of amino acids
    - Plant sources of amino acids are typically lacking one or more essential amino acid
  - Circulating amino acids are taken up by cells and used in biosynthesis
    - Liver deaminates amino acids; ammonia is formed
    - Ammonia is converted to urea and excreted
    - Keto acid may be converted to carbohydrates or lipids and stored or used as an energy source

Vitamins

Organic compounds needed in small amounts which must be supplied by the diet

Functions:

Various; many are coenzymes

Fat soluble vitamins: A, D, E, K

May be harmful if taken in megadoses

Water soluble vitamins: B and C vitamins

Minerals

Inorganic compounds needed in small amounts which must be supplied by the diet

Essential minerals needed in amounts of 100 mg or more daily:

Na, Cl, K, Ca, P, Mg, S

Trace elements are needed in smaller amounts, e.g. Fe, I

Iron deficiency is the most common mineral deficiency

Energy balance

Metabolic rate is the amount of energy utilized per unit time

BMR is the energy required just to be alive

When energy output exceeds intake, weight is lost

When appropriately balanced and managed - "a diet"

When energy output is less than intake, weight is gained

Obesity results from excess nutrients being stored as fat in adipose cells

Malnutrition is a result of lack of appropriate nutrients

Severe protein deficiency is known as kwashiorkor

Growth is stunted, abdominal edema is marked

Research and Discussion Topics

• Investigate the causes, symptoms and treatments of colon cancer. Where do most cancers of the colon occur? How do tests like the occult blood test, or sigmoidoscopy detect cancer of the colon?

Teaching Suggestions

• Students always find a description of kwashiorkor fascinating. It seems counter-intuitive that starving children, as often depicted on TV or the print media, have distended bellies, but wasted limbs. It is due to lack of protein, including the plasma proteins, and the osmotic balance of the blood is upset, and fluid pools in the abdomen. Kwashiorkor has also been advanced by the aggressive marketing by companies which make infant formula. They have given free samples to women in developing countries, and when making the switch to formula, the women stop lactating. After free samples are used up, they often "stretch" the formula, mixing it with increasing amounts of water, and the children suffer protein deficiencies. The childen may become sick from the use of contaminated water to make the formula, and also miss the immunological advantage of breast feeding. Fortunately, most major companies have stopped this marketing practice.

Suggested Readings

Sanderson, S.L. and R. Wassersug. "Suspension feeding vertebrates." *Scientific American*. March 1990. 96-101. Filter feeders, from whales to ducks.

Cohen, L.A. "Diet and Cancer." *Scientific American*. 1987. 257 (5) Correlations between dietary fat and cancers.

Chapter 40. Osmoregulation, Disposal of Metabolic Wastes, and Temperature Regulation

Chapter Overview

Metabolic wastes are excreted in animals via the skin, lungs or gills, and specialized excretory systems, such as kidneys. The skin and respiratory structures may excrete carbon dioxide and water, but nitrogenous wastes must be excreted by the gills or specialized structures. Ammonia is produced by protein metabolism, and in fish is excreted across the gills constantly. In other animals, ammonia is converted to uric acid, which is crystalline and can be excreted with little water loss (reptiles and mammals) or to urea, which is water soluble, seen in amphibians and mammals.

Most marine invertebrates are osmoconformers; invertebrates which live in estuarine or freshwater habitats are usuallyosmoregulaters. Terrestrial animals have excretory structures which conserve water. Specialized excretory structures in invertebrates include nephridia and Malpighian tubules. Vertebrates have kidneys for both osmoregulation and excretion.

Vertebrates evolved in freshwater, and the kidney of fish and amphibians excretes a large amount of dilute urine. Gills are the site of salt uptake in freshwater fish. Marine fish have the opposite situation, they drink water, and their gills excrete excess salts. The chondrichthyean fish maintain slightly hypertonic body fluids by accumulating urea in the tissues. Marine mammals produce a very concentrated urine due to the ingestion of salts from the water and in their food.

The mammalian urinary system is composed of the paired kidneys and ureters, the urinary bladder and the urethra. The kidneys are composed of millions of nephrons, which are the functional unit of the kidney. The glomerulus is the site of filtration, and the filtrate, which consists of water and small dissolved molecules passes into the Bowman's capsule. This large volume of filtrate is then processed in the renal tubules; materials are reabsorbed by active transport, diffusion and osmosis, and further modified by tubular secretion. This reabsorption results in a urine which has a high concentration of urea, and variable amounts of salts. The hormones ADH and aldosterone affect various sections of the nephron and regulate the volume and concentration of the urine.

Ectotherms have a body temperature that fluctuates with the environmental temperature. Behavioral strategies like basking may optimize body temperatures. This is energetically inexpensive but restricts the distribution of ectotherms. Birds and mammals are endothermic, and maintain a relatively constant, high body temperature. The hypothalamus and autonomic nervous system act to regulate body temperature. This is metabolically optimal, but energetically expensive.

Lecture Outline

Excretory systems

- Aid in maintenance of homeostasis
- Functions:
  - Osmoregulation; regulates water and salt balance
  - Excretion of metabolic wastes
    - Water
    - Carbon dioxide (lost via respiratory system)
    - Nitrogenous wastes
      - Ammonia produced during amino acid breakdown
        - Is toxic; must be excreted rapidly (fish)
        - Typically converted to uric acid or urea
      - Uric acid produced both from ammonia and breakdown of nucleic acids
        - Excreted as a paste; saves on water
        - Primary excretory product of birds, reptiles
      - Urea is primary excretory product of amphibians and mammals
        - Produced in liver
        - Dissolved in water= urine

Invertebrate excretory structures

- Most marine invertebrates are osmotic conformers
  - Osmotic regulators are seen in areas like estuaries, where salinity varies
- Nephridia are tubules which open to the exterior of the body
  - Flatworms have protonephridia
  - Annelids and molluscs have metanephridia
    - Coelomic fluid flows into tubules
    - Some materials are selectively reabsorbed
    - Relatively concentrated urine is produced
- Malpighian tubules are seen in insects and arachnids
  - Tubules collect coelomic wastes, emptied into the intestine
  - Rectal glands reabsorb water, some salts
  - Concentrated paste of uric acid is produced

Osmoregulation in vertebrates

- Primary osmoregulatory and excretory organ in vertebrates
  - Supplemented by skin, lungs or gills, digestive system, specialized salt glands
- Osmoregulation
  - First freshwater vertebrates (fish) moved into hypotonic environment
    - Scales and mucus slows influx of water
      - Gills are no impediment to water inflow
      - Kidneys excrete a copious dilute urine

- Salts are lost
    - Chloride cells in gills transport salt into the body
- Amphibians have similar mechanisms
    - Salt is transported into body via cutaneous active transport
- Marine bony fish have body fluids hypotonic to salt water
    - Tend to lose fluid, so drink seawater
    - Excrete excess salts by the same chloride cells in gills
    - Kidneys excrete very little urine; have reduced glomeruli
- Marine chondrichthyean fish concentrate urea in tissues
    - Tissues are therefore slightly hypertonic to sea water
    - Kidneys excrete a large volume of urine
    - Rectal gland also excretes salt
- Marine mammals ingest salt water while eating
    - Kidneys produce a concentrated urine

<u>The mammalian excretory system</u>

- Excretory organs
    - Carbon dioxide and water excreted by lungs
    - Sweat glands in skin excrete salt
    - Bile pigments excreted by liver into the intestine
        - Liver is also the site of production of urea and uric acid
- Kidneys
    - Paired, dorsally located in upper lumbar region
    - Outer portion, the cortex
    - Inner portion, the medulla
    - Urine flows into the renal pelvis, then to the ureters
    - Ureters connect to the urinary bladder
    - During urination, urine flows through the urethra to the outside
        - In males, the urethra carries urine or semen
        - Bladder infections are more common in women, as the urethra is very short

<u>The nephron</u>

- The functional unit of the kidney
- The Bowman's capsule (glomerular capsule) collects filtrate, connected to the renal tubules
- The glomerulus is the capillary bed surrounded by the Bowman's capsule
- The afferent arteriole feeds the glomerulus; the efferent arteriole drains it
    - Due to blood pressure, some plasma is forced out of the capillaries
- Blood from the efferent arteriole passes into a second set of capillaries
    - These capillaries surround the renal tubules
    - These capillaries pick up reabsorbed materials
    - Ultimately, drain into the renal vein
- Filtration of materials from the glomerulus to the Bowman's capsule
    - Forced by hydrostatic pressure; afferent arteriole is larger than efferent
    - Glomerular capillaries are highly permeable

Filtrate contains:
- Water; incredible volumes! 45 gallons per day
- Small dissolved molecules; salts, urea, glucose
- Blood cells remain in the capillaries
- Large molecules like proteins remain in the capillaries

Most of the filtrate is reabsorbed in the renal tubules
- Most of the water, glucose, other small molecules reabsorbed
- Accomplished by active transport, diffusion and osmosis

Renal threshold
- The maximum concentration of a substance that the kidney can reabsorb
- In uncontrolled diabetes mellitus, glucose is present in the filtrate in excess of the renal threshold; excess is lost in the urine

Some materials are actively secreted in the distal portion of the renal tubules
- Potassium, hydrogen, and ammonium ions as well as drugs, such as penicillin are secreted into the filtrate

Urine is concentrated in the renal tubule
- Sodium ions are actively transported out of the proximal tubule
  - Water follows osmotically
- Walls of the descending loop of Henle (loop of the nephron) are impermeable to sodium and urea, but is permeable to water
  - Interstitial fluid is high in salt
  - Water moves out of the descending loop by osmosis
- Walls of the ascending loop of Henle are permeable to salt, less permeable to water
  - Salt diffuses out in lower portions, actively transported out in upper portions
  - This maintains the high external salt concentration needed for the mechanism of the descending loop
  - The longer the loop, the more concentrated the urine
- Collecting tubules carry urine from many nephrons
  - As they pass through the area of high salt concentration, more water is lost by osmosis

Urine production is regulated by ADH; Antidiuretic hormone
- Produced by the posterior pituitary gland
- Causes collecting ducts to be more permeable to water
  - Results in a more concentrated urine
- Diabetes insipidus results from insufficient ADH
  - Production of copious dilute urine
  - Controlled by injections of ADH or nasal spray

Aldosterone controls sodium reabsorption
- Produced by the adrenal cortex
- Simulates the distal tubules and collecting ducts to increase reabsorption of sodium
- Stimulated by a decrease in blood pressure
  - Kidney cells secrete renin

Stimulates the renin-angiotensin pathway
Renin converts a plasma protein to angiotensin
Angiotensin stimulates aldosterone production
Also stimulates smooth muscles in arteriole walls
Urine is the final product which enters the renal pelvis
96% water, 2.5% urea, 1.5% salts, trace amounts of other materials
Urinalysis is the examination of urine

Temperature regulation
Ectotherms are "cold blooded"; body temperature is derived from heat from the environment
Tend to have fluctuating body temperatures
Some ectotherms can produce their own heat by muscular activity
Advantage: less energy cost
Endotherms maintain a relatively constant temperature, body temperature is derived from heat of metabolism
Temperature varies with muscular activity, hormonal action, sweating
The hypothalamus and thermoreceptors in the skin sense heat
Autonomic controls on dilation/constriction of blood vessels
Advantage: increased metabolic activity

Research and Discussion Topics

• Compare the functioning of the kidneys of a desert mammal, a human and a freshwater mammal, like the beaver. What would be the expected variation in concentration of urine produced? What about the length of the nephron tubules?

• The kangaroo rat has a variety of adaptations to allow it to live in a habitat with minimal water. Research the physiological adaptations of this little rodent.

• Some current research indicates that the dinosaurs may have been inertial homeotherms, or actually endotherms. Investigate the evidence for these hypotheses.

Teaching Suggestions

• Students often have a hard time keeping all of these osmoregulatory mechanisms straight. I typically draw an outline of the organism on the board, then add arrows showing major fluxes of water and salts, and the type of urine produced. For example, draw an outline of a freshwater fish, show influx of water, production of copious urine. Compare and contrast to a marine fish, and a marine mammal. Also compare and contrast a marine mammal which eats invertebrates, which are typically osmoconformers, or a marine mammal which eats other vertebrates, whose tissues are typically hypotonic to the sea water. Which would have the most problem with salt balance?

Chapter 41. Endocrine Regulation

Chapter Overview

Endocrine glands produce hormones, which travel primarily in the circulatory system to affect target organs. Hormone secretion is regulated by negative feedback mechanisms, and is further influenced by the production of hypothalamic releasing and release-inhibiting factors, and pituitary tropic hormones. Hormones enter cells directly, or indirectly, by binding with receptors in the plasma membrane. cAMP is a well known second messenger.

Invertebrate hormones are typically neurohormones, and affect development and behavior. Vertebrate hormones are typically produced by endocrine tissues, and are of clinical interest when hypersecreted or hyposecreted.

The posterior pituitary secretes hormones produced by neurons in the hypothalamus; oxytocin and ADH. The anterior pituitary is glandular, and produces prolactin, growth hormone and several tropic hormones. Release of these hormones is influenced by chemicals from the hypothalamus via a portal connection, as well as exercise, sleep and social interactions. Growth hormone stimulates cells to increase protein synthesis.

The thyroid gland produces thyroid hormone, which stimulates metabolism, and calcitonin, which is involved in calcium levels. Antagonistic in action to calcitonin, parathyroid hormone is also involved in calcium balance. The pancreas is endocrine, as well as exocrine, and secretes insulin and glucagon. These two hormones act antagonistically, and in tandem to maintain a relatively constant blood sugar level.

The adrenal medulla is an endocrine gland which is stimulated by the sympathetic division of the autonomic nervous system, and the action of its hormones is similar to sympathetic stimulation, but has a longer lasting effect. The adrenal cortex produces corticosteroids; the androgens, mineralocorticoids and glucocorticoids. These have wide-ranging effects on salt balance and nutrient metabolism. Many other endocrine tissues are known, including the heart, the kidney and the digestive tract.

Lecture Outline

<u>Hormones</u>

- Definition:
  - Chemical messengers which regulate the activity of other tissues
- Hormones are produced by glands of the endocrine system
  - Hormones are secreted into the surrounding tissue fluid, then
    - diffuse into the blood; are transported by the blood
  - Stimulate only target tissues

Hormones are either lipids or proteins
Now know that nearly all tissues of the body are endocrine

Hormone activity

Negative-feedback mechanisms
- Effects are opposite of the stimulus
- When end-product concentration rises, hormone production stops
- When end-product concentration drops, hormone production resumes

Hormones only affect the target tissue
- Hormones bind to a receptor in target tissues

Small polypeptide molecules pass through the plasma membrane
- Combine with receptors in the cytoplasm or nucleus
- Hormone-receptor complex acts with proteins in the nucleus to activate genes coding for certain proteins

Other hormones combine with receptors on the plasma membrane of the target cell
- The message is relayed in the cell by the second messenger
  - cAMP (cyclic AMP) is much studied
  - Hormone binds with enzyme, enzyme catalyzes ATP to form cAMP
  - cAMP activates enzymes which are themselves inhibitory or stimulatory

Prostaglandins are local hormones; act on nearby cells
- Modified fatty acids; produced by many organs
- Mimic actions of cAMP, interact with other hormones
- Approximately 16 are known
- Varying functions; much interest in clinical use

Invertebrate hormones

Primarily produced by neurons; called neurohormones
- Affect many developmental changes, behavior

Vertebrate hormones

Various organs, functions, but tend to be similar among vertebrates

Endocrine disorders
- Hypersecretion hyperstimulates target cells
- Hyposecretion hypostimulates target cells
- Some disorders involve defects in receptors

Endocrine regulation by the hypothalamus
- A link between the nervous and endocrine systems
- Physically connected to the pituitary gland by a portal system
- Secretes releasing and release-inhibiting hormones

Posterior lobe of the pituitary gland

Hormones are produced by nerve cells in the hypothalamus
- Flow down axons from hypothalamus to posterior pituitary
- When neuron is stimulated, vesicles are released

Oxytocin
Causes uterine contractions during childbirth, contraction of smooth muscles in the breast for milk expulsion
ADH
Targets cells in the nephron
Causes a lesser volume of urine to be produced

Anterior lobe of the pituitary gland
Prolactin
Stimulates milk production by mammary tissue
Tropic hormones; hormones which stimulate other endocrine glands
TSH
Thyroid stimulating hormone; targets the thyroid gland
ACTH
Adrenocorticotropic hormone; targets the adrenal cortex
FSH, LH
Gonadotropic hormones; follicle-stimulating hormone, leutenizing hormone
Human growth hormone
Stimulates growth by increasing protein synthesis
Mobilizes fat from adipose tissues
Secretion affected by inhibiting and stimulatory factors
Also affected by exercise, non-REM sleep, social interactions, other hormones
Clinical application:
Pituitary dwarfs are a result of juvenile hyposecretion
Gigantism is a result of juvenile hypersecretion
Circulating levels in adults are very low; no problems with adult hyposecretion
Acromegaly results from adult hypersecretion; abnormal growth of the hands, face

Thyroid gland
Located in the neck, below the larynx
Thyroid hormone
Thyroxine (T_4) and triiodothyronine (T_3) are made from the amino acid tyrosine and iodine
Functions in growth; stimulates metabolic rate
Important in development, for example amphibian metamorphosis
Regulated by negative-feedback hormones
Also affected by the hypothalamus, TSH from the pituitary
Acts by way of cyclic AMP
Clinical applications:
Juvenile hyposecretion leads to cretinism
Adult hyposecretion causes myxedema
Adult hypersecretion causes Grave's disease
Very rapid metabolic rate

- A goiter is an enlarged thyroid
  - Often caused by iodine deficiency; no hormone production, TSH levels rise, thyroid enlarges
- Calcitonin
  - Inhibits removal of calcium from bone when plasma calcium levels rise

<u>Parathyroid glands</u>

- Embedded in the posterior surface of the thyroid
- Parathyroid hormone
  - Functions in releasing calcium from bones when plasma calcium levels drop
  - Also activates vitamin D, which aids in intestinal calcium uptake
  - Acts with calcitonin to maintain steady plasma calcium levels

<u>Pancreas</u>

- Endocrine tissue is located in the pancreatic islets, the islets of Langerhans
- Alpha cells produce insulin
  - Insulin stimulates cells to take up glucose and other nutrients
- Beta cells produce glucagon
  - Stimulates liver cells to release glucose and other nutrients
  - Action is opposite of insulin
- Both hormones are regulated by sugar concentration, act together to maintain relatively constant blood glucose levels
- Diabetes mellitus is the most common endocrine disorder
  - Type I (insulin-dependent diabetes); early onset
    - Low numbers of beta cells in pancreas; may be an autoimmune condition
    - Require insulin injections
  - Type II (non-insulin dependent diabetes); later onset
    - Lack insulin receptors on cells
  - Both may experience hyperglycemia, rapid fat mobilization, which leads to atherosclerosis, and increased protein breakdown

<u>Adrenal (suprarenal) gland</u>

- Adrenal medulla
  - Inner portion of the gland
  - Secretes epinephrine (adrenaline) and norepinephrine
  - Stimulated by the sympathetic nervous system
    - Longer lasting effect than sympathetic nervous stimulation alone
    - Same effects as sympathetic stimulation
- Adrenal cortex
  - Outer portion
  - Produces steroid hormones (made from cholesterol)
  - Androgens
    - Have masculinizing effects

Mineralocorticoids
- Aldosterone is the primary mineralocorticoid
- Stimulate nephron to reabsorb sodium, excrete potassium
- Hyposecretion results in water loss, blood pressure drops

Glucocorticoids
- Cortisol (hydrocortisone) is the principal glucocorticoid
- Stimulates the liver to produce glucose and glycogen
- Stress causes the hypothalamus to secrete CRF (corticotropin releasing factor), which stimulates the pituitary to release ACTH, which stimulates the adrenal cortex to produce glucocorticoids

Clinical applications
- Hyposecretion of cortical hormones causes Addison's disease
- Hypersecretion of cortical hormones causes Cushing's disease
- Glucocorticoids are used to reduce inflammation, allergies

Other endocrine tissues

- Many hormones secreted by the digestive tract regulate digestive functions
- The thymus produces thymosin; important in immunity
- The kidneys secrete renin and erythropoetin
- The heart secretes ANF (atrial natriuretic factor)
  - Inhibits aldosterone and renin release
- The pineal gland produces melatonin

Research and Discussion Topics

• Research the myriad of effects that anabolic steroids have on users. What links to cancers have been shown?

• Genetically engineered human growth hormone is now being used to treat pituitary dwarfs. Describe its other uses, such as in treatment of burn patients, and AIDS patients.

Teaching Suggestions

• Students are typically mesmerized by pictures of endocrine disorders. I usually xerox and make transparencies of persons with the conditions of hypo- and hypersecretions. I find particularly interesting progressive conditions like acromegaly, in which the person looks perfectly "normal", and then photos show the progressive course of the disease.

Suggested Readings

Hanson, B. "Raging hormones at the White House." *Discover*. January 1992. 54. Grave's disease affects George and Barbara Bush, and their dog.

Diamond, J. "Turning a man." *Discover*. June 1992. 71-77.
and
Grady, D. "Sex test of champions." *Discover*. June 1992. 78-82. Two very compelling articles discussing pseudohermaphrodites, the related genetic testing of female athletes, and social acceptance of pseudohermaphrodites in different cultures.

Clark, M, D. Gelman, M. Gosnell, M. Hager and B. Schuler. "A users guide to hormones." *Newsweek*. January 12, 1987. 50-59. Effects of hormones on growth, sex and behavior.

Chapter 42. Reproduction

Chapter Overview

Asexual reproduction results in offspring identical to the parent; sexual reproduction, which involves fusion of gametes results in variation in offspring. Most animals are either male (and produce sperm), or female (and produce eggs); hermaphrodites produce both.

In males, sperm are produced in the seminiferous tubules of the testes. Between the tubules are interstitial cells which produce testosterone. Sperm finally mature in the epididymis, where they are ultimately ejaculated or absorbed. Ejaculated sperm travel up through the inguinal canal, passing into the ejaculatory duct, which joins the urethra in the center of the prostate. Three accessory sex organs contribute fluids to the sperm, making semen. The semen in the urethra then travels out of the penis. The erectile tissue of the penis maintains erection during sexual intercourse, and facilitates sperm delivery into the vagina of the female.

Females have sexual structures homologous to the male reproductive structures. Eggs are produced in the ovaries, which do not descend as the testes do, but remain up in the pelvic cavity, suspended by ligaments. Eggs begin meiosis before birth, but remain in a state of arrested development. After puberty, under the influence of hormones, one or several eggs begin further development in the follicle. At ovulation, the follicle ruptures, and the egg may pass into the uterine tube. The tissues of the spent follicle become the corpus luteum, which secretes hormones for several weeks, or longer if fertilization occurs. If fertilized, the egg passes down the uterine tubes to the uterus where it implants. If fertilization does not occur, hormonal changes cause the endometrium of the uterus to be shed, in menstruation.

The vulva is the collection of external structures; the labia majora and minora, and the clitoris, which are homologous to the scrotum and the penis, respectively. The female breasts function in lactation; milk is produced in the alveoli.

Male and female gamete production varies; males produce 4 sperm from 4 primary spermatocytes. Females produce 1 egg and a variable number of polar bodies from 1 primary oocyte. Eggs are large, have much cytoplasm, and are nonmotile. Sperm are small, have little cytoplasm and are motile. Fertilization of the egg by one sperm restores the diploid number of chromosomes. A successful fertilization takes place in the uterine tubes, followed by implantation in the thickened endometrium.

Birth control mechanisms include hormonal means which mimic pregnancy and ovulation does not occur, and mechanical methods which prevent sperm from entering the female's body or cervix (condom or diaphragm respectively). An IUD prevents implantation of a fertilized egg. Sterilization prevents passage of the sperm or egg through the vas deferens or uterine tube. Abortions may be non-induced (miscarriages) or induced to terminate the pregnancy.

Lecture Outline

Asexual reproduction

- A single parent gives rise to offspring which are genetically identical
  - Does not involve gametes
- Sponges and cnidarians may reproduce by budding
- Some echinoderms reproduce by fragmentation
- Parthenogenesis is common in molluscs, some crustaceans, insects, and some reptiles
  - Unfertilized eggs develop into adults
  - Males are produced at certain times of the year

Sexual reproduction

- Involves fusion of gametes
- Male parent produces sperm, female parent produces eggs
  - Eggs are typically large and nonmotile, sperm are small and motile
- External fertilization results from mutual shedding of gametes
- Internal fertilization typically involves an intromittent organ
  - Optimal because eggs can develop within the female, or have a shell
- Hermaphrodites produce both eggs and sperm in one animal
  - May self fertilize, or cross fertilize by mutual exchange of sperm

Human male reproductive system

- Testes
  - Site of spermatogenesis; in the seminiferous tubules
  - Spermatogonia divide by mitosis; forming primary spermatocytes
  - Primary spermatocyte divide meiotically to produce spermatids which are haploid
  - Sperm develop in the testes and epididymis
    - Located in the scrotum; optimal lower temperature
  - Mature sperm have an enzyme cap, a flagellum, and little cytoplasm
- Epididymis
  - Final maturation of sperm
  - Are ejaculated from here
  - Reabsorbed if not ejaculated
- The inguinal canal
  - Carries the vas deferens, nerves, vessels, muscles between the abdomen and testes
  - A weak spot; an inguinal hernia may occur here
- Vas deferens (ductus deferens)
  - Muscular tube which carries sperm from epididymis up into the pelvic cavity
  - Ends in the ejaculatory duct, which empties into the urethra in the center of the prostate
- The urethra carries both urine and semen, but not simultaneously
- Accessory glands produce semen
  - Prostate gland produces an alkaline secretion

- Paired seminal vesicles produce a fluid high in fructose
- Bulbourethral glands secrete mucus

The penis contains columns of erectile tissue
- During erection, arteries dilate, the erectile columns swell with blood
- Swelling compresses the veins, slowing venous drainage, maintaining the erection
- Bacula (penis bones) seen in some mammals (not humans!)

Reproductive hormones
- Promote both male primary sexual characteristics (in fetal development) and secondary sexual characteristics (at puberty)
- GnRH (gonadotropin-releasing hormone) stimulates the anterior pituitary to produce FSH and LH
- FSH stimulates development of seminiferous tubules
- LH stimulates the interstitial cells to secrete testosterone

Human female reproductive system

Ovaries
- Are homologous to the testes
- Produce both eggs and hormones
- Suspended in the pelvic cavity by ligaments of connective tissue
- Ova are in varying stages of arrested meiotic divisions
  - Primary oocytes enter 1st meiotic division by the time of birth
  - At puberty, one or several resume development each month
  - Primary oocyte is surrounded by cells; the follicle
    - Primary oocyte--> 1 polar body and a secondary oocyte
    - The secondary oocyte enters meiosis II
      - Completes meiosis after fertilization
  - Hormones cause follicular maturation
  - After ovulation, corpus luteum secretes estrogens and progesterone

Oviducts (or uterine tubes, previously called the Fallopian tube)
- Lined with cilia, propels the egg toward the uterus
- If fertilization is successful, it happens here

Uterus
- Muscular organ which incubates the embryo
- Embryo implants here, develops placental attachment with uterine wall
- If no fertilization occurs, the lining (endometrium) sloughs off
- Endometriosis is the migration and growth of endometrial tissue elsewhere in the reproductive tract, or the abdomen
- During childbirth, the muscular wall contracts
- The cervix is the neck of the uterus
  - Common site for cancer; Pap test detects growth

The vagina
- The site of sperm ejaculation

Vulva
- All of the external structures surrounding the vaginal orifice

Labia majora and minor are the folds; homologous to the scrotum
Clitoris is homologous to the penis; is erectile, many nerve endings

Breasts (mammary glands)
- Function in milk production
- Lobes are glandular, surrounded by a variable amount of adipose tissue
- The alveolar cells produce milk, flows through ducts, ultimately through the opening in the nipple
- A common site of cancer

Reproductive hormones
- Estrogens are responsible for secondary sex characteristics
- Menstrual cycle regulated by hormones from hypothalamus, anterior pituitary and ovaries
- Menstruation is the shedding of the uterine endometrium
- During the preovulatory phase, follicles secrete estrogens
    - The endometrium begins to thicken
    - Estrogens stimulates the anterior pituitary to produce LH
    - LH and FSH cause final maturation of the follicle
- Ovulation follows 14 days after the onset of menstruation
- The postovulatory phase is marked by development of the corpus luteum, which produces progesterone and estrogens
- If no fertilization occurs, the corpus luteum degenerates
- If the egg is fertilized, it implants in the endometrium, and secretes hCG (human chorionic gonadotropin)
- hCG signals the corpus luteum to continue to produce hormones
- hCG is the basis of the pregnancy test

<u>Sexual intercourse</u>
- Semen is ejected into the vagina
- Involves 4 phases: sexual desire, excitement, orgasm and resolution
    - Desire is self explanatory!
    - Excitement involves vasocongestion and muscle tension
    - Orgasm involves rhythmic contraction resulting in ejaculation
    - Resolution follows orgasm; a refractory period

<u>Fertilization</u>
- Functions:
    - Restores the diploid number of chromosomes
    - Determines sex of offspring in most animals (XX female, XY male)
    - Egg completes meiosis
- Conception includes fertilization and establishment of pregnancy
- Sperm swim up to the oviduct within 5 minutes
    - Many sperm must surround the egg for fertilization to occur
    - Sperm acrosome recognizes proteins on the egg
    - Sperm enter the egg, followed by a rapid electrical change
        - Blocks polyspermy
    - The pronuclei fuse forming the diploid zygote nucleus

Birth control

- Birth control mechanisms may prevent fertilization, or prevent conception
- Hormone contraceptives prevent ovulation
  - Mimic pregnancy, so body does not ovulate
  - Are synthetic hormones
  - Contraindicated for women who smoke or have hypertension
  - Depo-provera and Norplant are newer forms
- IUD's prevent implantation
  - Disadvantages include PID, complications if pregnancy does occur
- Diaphragm
  - Prevents sperm passage into the cervix
  - Should be used with spermicides
- Condoms
  - Mechanical barrier; also prevent transmission of STD's
- Sterilization is very reliable; most are done on males
  - In a vasectomy, the vas deferens is cut
    - May be reversible
  - Tubal ligations involve cutting the uterine tubes
- Abortion is the termination of a pregnancy
  - Spontaneous abortions (miscarriages) occur "naturally"
  - Theraputic abortions are performed to maintain the health of the mother, or in the case of birth defects
  - Abortion for birth control is very controversial

Sexually transmitted diseases

- The second most common communicable diseases in the world, after the cold
- Gonorrhea affects 250 million people world wide
- Most common STD in the US is chlamydia
- AIDS is a rapidly rising STD

Research and Discussion Topics

• Research the various chemical/hormonal options in contraception. How do they work? What are the pros and cons of each method?

• Discuss the homology of the male and female sexual structures. Trace their development in the embryo, from the gender indifferent to the distinguishable male and female embryos.

• Thinking about the timing of production of gametes, why is the age of the mother linked to more genetic disorders than the age of the male?

Teaching Suggestions

• I don't think I've ever had so much student interest in any subject except sex, and particularly relating to sexual abnormalities. I discuss the anomalies which result in the "pseudohermaphrodite" condition, in which hormonal abnormalities result in variable gender conditions. We discuss the testing of female athletes (to make sure they are genetically female), and tie in the effects of hormones on development of sexual structures.

• Students may find it interesting to know that male hormones cycle as well, but are on a more annual cycle, with the peak in spring (no surprise!). It has been shown that male students do better on tests when their testosterone levels are higher (don't tell your fall semester students that). It has also been shown that lawyers, yes, do have higher levels of testosterone than men in other occupations.

• The interplay between the female hormones and the stages of the menstrual cycle can be confusing. I try to explain to them that after ovulation, the body makes changes that get it ready for pregnancy, if that should occur. Uterine and breast tissue proliferates, etc. Then, if implantation does not occur, the corpus luteum degenerates, progesterone levels drop, and the endometrium is shed. Personally, I would have "designed" females the other way around--make the appropriate changes only if fertilization occurs!

Suggested Readings

Diamond, J. "Turning a man." *Discover*. June 1992. 71-77.
and
Grady, D. "Sex test of champions." *Discover*. June 1992. 78-82. Two very compelling articles discussing pseudohermaphrodites, testing of athletes, and social acceptance.

Anomymous, "Testy fellows." *Science*. 26 April 1991. 513. Testosterone levels are highest in lawyers and actors.

Freedman, D.H. "The aggressive egg." *Discover*. June 1992. 61-65.
and
Ansley, D. "Sperm tails." *Discover*. June 1992. 66-69. Two articles which analyze the varying factors involved in fertilization.

Wassarman, P.M. "The biology and chemistry of fertilization." *Science*. 30 January 1987. 553-560. The molecular basis of mamalian fertilization.

Aral, S.O. and K.K. Holmes. "Sexually transmitted diseases in the AIDS era." *Scientific American*. February 1991. 62-69. Syphilis, gonorrhea and other STD's in today's world.

Nimmons, D. "Sex and the brain." *Discover*. March 1994. 64-71.
and
Le Vay, S. "A difference in hypothalamic structure between heterosexual and homosexual men. *Science*. August 1991. 253: 1034-1036. (3 other articles in this issue as well). A description of Simon Le Vay's research on differences between the brains of homosexuals and heterosexuals.

Wallace, R.A. "How they do it." 1980. Morrow Quill Press. 124 pages. A humorous, yet biologically based account of sex in the animal world.

Chapter 43. Development

Chapter Overview

Development of an animal begins shortly after fertilization, and cells divide mitotically, forming first a morula, then a hollow blastula. Ultimately, the blastula undergoes an infolding to form the gastrula. The cavity inside the gastrula is known as the archenteron, which will ultimately become the digestive cavity.

Cell divisions may be equal or unequal; unequal typically seen in eggs with much yolk. The cells which contain the yolk (the vegetal pole) divide slowly; the cells at the other end, the animal pole, are more active, and divide rapidly. Radial cleavage is characteristic of deuterostomes, spiral cleavage is characteristic of protostomes. Further, the opening to the archenteron, the blastopore, becomes the mouth in protostomes, and the anus in the deuterostomes. At this time, the embryo has three cell layers; the endoderm which forms the digestive tract, the ectoderm, which makes up the outer covering of the embryo, and gives rise to the skin and nervous system, and the middle mesoderm, which ultimately forms most of the internal non-digestive organs.

The nervous system is one of the first to develop, and forms from invagination of ectoderm, forming a dorsal hollow structure. The anterior portion forms the brain, the posterior portion forms the spinal cord.

Reptiles, birds and mammals have 4 extraembryonic membranes; the chorion and amnion, which surround and protect the embryo; the yolk sac which contains the yolk; the the allantois, which stores wastes or may be erythropoetic.

Human development begins in the uterine tubes; after fertilization begins cleavage, and the ball of cells travels to the uterus where it ultimately implants by the enzymatic action of the trophoblast cells. The placenta is the site of exchange between fetus and maternal tissues; specifically the chorionic villus and the endometrial cells.

The most critical developmental stages of the human embryo occur during the first trimester. During this time, all organ systems begin development. During the second and third trimesters, the fetus grows rapidly, and moves often.

Birth is characterized by contraction of the uterine muscle and dilation of the cervix. After birth of the baby, the placenta and extraembryonic membranes are expelled.

Further developmental changes occur throughout the life of the person, becoming marked in old age, when the body does not maintain homeostasis as efficiently as before, and various organ systems function less efficiently.

Lecture Outline

Overview of development

- Development includes all of changes from fertilization to death
- Cell division is first event
- Morphogenesis includes differentiation, controlled cell migration and death to form tissues and organs

Cleavage

- Fertilized egg divided mitotically to form 2,4,8,16 cell zygote
- Ball of cells is called the morula
- The hollow ball stage is the blastula (cavity is blastocoel)
- Cell division is not equal
  - Some cells have a greater proportion of cytoplasm
  - In amphibians, the gray crescent forms
    - Is opposite the side where the sperm penetrated
    - The gray crescent ultimately becomes the dorsum of the embryo
  - Vertebrate eggs are polarized
    - In eggs with much yolk, the yolky end is the vegetal pole
    - The opposite end (more metabolically active) is the animal pole
      - Animal pole is primary site of cleavage
      - Vegetal pole is relatively slower
      - Cleavage results in the blastocoel displaced upward
  - Cleavage patterns
    - In eggs with little yolk, or evenly distributed yolk, radial (more even) cleavage occurs
      - Cleavage occurs at right angles to previous cleavage
      - Characteristic of deuterostomes (echinoderms and chordates)
    - In eggs with much yolk, see spiral cleavage
      - Cleavage is oblique; cells end up being spirally arranged
      - Characteristic of protostomes

Germ layers

- Blastula becomes gastrula by infolding (invagination)
  - This obliterates the blastocoel
- Forms three embryonic cell layers
  - Endoderm lines the archenteron (inner cavity)
    - Archenteron ultimately becomes the gut
  - Ectoderm is the outer layer of cells
    - Develops into the skin, nervous and sensory structures
  - Mesoderm develops between endoderm and ectoderm
    - Ultimately forms skeletal, muscular, circulatory, excretory and reproductive system

In deuterostomes, the opening to the archenteron becomes the anus
Another opening later develops into the mouth
In protostomes, the opening to the archenteron becomes the mouth

Early organogenesis
In vertebrates, the notochord forms on the dorsal surface
The notochord induces overlying ectoderm to form the neural plate
The neural plate folds in forming the neural groove
The edges of the groove are the neural folds
The two neural folds move together and fuse, thus forming the neural tube
The neural tube is thus formed by ectoderm; and is hollow
The anterior portion of the neural tube forms the brain
The posterior portion of the neural tube forms the spinal cord

Extraembryonic membranes of reptilian, avian and mammalian vertebrates
4 membranes; are not part of the embryo and are separated at birth
Are adaptations to terrestrial life
Chorion and amnion
Surround the embryo with a cushion of fluid
Allantois
An outgrowth of the digestive tract
In reptiles and birds, stores nitrogenous wastes
Reduced in mammals
Contributes to the formation of the umbilical circulation
Yolk sac
In animals with much yolk, serves as a nutritive supply
In animals with little yolk, site of blood cell formation

Human embryonic development
Gestation period is approximately 266 days from conception
Development first begins in the uterine tube
First division occurs about 24 hours after fertilization
Cleavage occurs as zygote moves down uterine tube
Cilia and peristalsis aid in movement
Embryo reaches uterus after 5 days, floats free for several days, nourished by uterine secretions
Is a blastocyst at this stage (the term for mammalian blastulas)
Outer cells, the trophoblast, will form the chorion and amnion
The inner cell mass forms the embryo itself

Implantation and placental development
Implants in endometrium by about 7th day
Cells of trophoblast enzymatically erode a patch of endometrium
Embryo moves into the thick endometrium, develops there
The placenta in mammals is the site of maternal-fetal exchange
Develops from embryonic chorion and maternal endometrium
Chorionic villi become vascularized, grow into endometrium

Umbilical cord connects embryo and placenta
Two umbilical arteries connect to the capillaries of the villi
One umbilical vein returns blood to the embryo
Exchanges of materials take place in the placenta
Exchanges occur by diffusion or active transport
Maternal and fetal blood does not mix (the contrary is a very common misconception)
Trophoblast of embryo produces HCG (human chorionic gonadotropin) after implantation
Signals corpus luteum to continue producing progesterone and estrogens
Hormones function in maintaining pregnancy

The first trimester of human development
Gastrulation occurs during the second and third development
Heartbeat may be detected by three and a half weeks
Brain development begins by the fifth week
Basic chordate characteristics seen; pharyngeal pouches, postanal tail
Respiratory system develops from base of pharynx; digestive organs differentiate
Limb buds grow, skeletal muscle allows some movement
By end of first 2 months, called a fetus- is slightly over 2" long
Sexual differentiation has taken place

The second and third trimesters of human development
Can hear fetal heart with a stethoscope
Rapid growth, final differentiation of organs
At birth, averages 7 lb.
If born before 37 weeks' gestation, is considered premature, often not viable

Birth, or parturition
Hormonal influence of oxytocin initiate labor
First stage
Lasts 12 hours
Uterine contractions move fetus towards cervix
Cervix dilates
Amniotic ruptures, fluid is lost
Second stage
20 minutes to an hour
Fetus is delivered by contractions of uterine muscles (smooth muscle) and skeletal muscles of the abdominal wall
Umbilical cord is tied and cut
Baby (neonate) breathes soon after birth
Initiated by accumulation of CO_2 in fetal blood, stimulates respiratory centers in medulla

Third stage
    10-15 minutes after birth of child
    Placenta and adjoining membranes are expelled
Labor may be induced by drugs
A Caesarean section may be done to deliver the baby through an incision in the abdominal wall

Environmental prenatal effects

Many drugs may affect fetal development
Approximately 5% of babies born in the US have a clinical birth defect
    May be genetic in cause
    May be due to environmental factors
        Fetal alcohol syndrome causes many cases of mental retardation
        Drugs like cocaine and heroin results in addicted babies
        Smoking causes babies to be born with lower-than-average birth weights
        HIV and syphilis may be transmitted to babies in utero, during birth and postpartum through breastfeeding (HIV only)
Embryos are most susceptible to environmental factors during first trimester
Amniocentesis and chorionic villus sampling can detect some defects
Sonograms may aid in visualization of the fetus

Aging

Many organ systems decline in function as we age
Varying systems decline; cells are lost, circulation is impaired
Ability to maintain homeostasis decreases
Of increasing importance as we are seeing a larger elderly population
Changes with age may be due to wear and tear, accumulation of toxins, accumulated exposure to X-rays etc.

Research and Discussion Topics

• Both drug-addicted babies and children affected by fetal alcohol syndrome (FAS) are an important health care issue. Investigate the ways these drugs affect fetal development, and what long-term developmental problems may occur. How are these neonates treated clinically?

• Vitamins are critical for the developing embryo, particularly in the first trimester. For example, investigate the correlation between low levels of folic acid and spinal tube defects, such as spina bifida. Do you agree with the statement that all women who might possibly become pregnant should take folic acid supplements? How many birth defects of this type might be avoided if women took these supplements?

Teaching Suggestions

• It is important to emphasize to students the importance of the nutrition and other habits of the pregnant female to the health of the developing embryo. Most students do not know that the first trimester is the critical time of development, and that the embryo is the most susceptible to the effects of drugs etc. in this period. Emphasize that someone who is sexually active and is not practicing contraception may become pregnant and unknowingly harm the fetus.

Suggested Readings

Gold, M. "The baby makers." *Science '85*. April 1985. 26-28. In vitro techniques.

Steinmetz, G. "Fetal alcohol syndrome." *National Geographic*. February 1992. 37-39.

Chapter 44. Animal Behavior

Chapter Overview

Behaviors are all of the responses of an organism to its environment, including other members of its species and other species. A simple behavior includes rhythmic behavior, which may be diurnal or lunar in nature. These rhythms are controlled by cellular processes; as well as the hypothalamus and pineal gland in higher vertebrates.

All behaviors have a genetic component. Some behaviors are completely innate, including fixed action patterns, which result from presentation of a sign stimulus. Many behaviors are influenced by learning. In classical conditioning, associations are made by the animal between a stimulus and a normal response. In operant conditioning, an outcome is associated with a learned response. Imprinting is an important learned response which develops just before or right after birth or hatching. The most complex behaviors involve insight learning, which is a hallmark of the primates.

Behavioral ecology includes the study of behaviors such as migrations, which are regular, seasonal movements of animals, typically between an overwintering area and breeding grounds. Optimal foraging is another social behavior which allows an animal to maximize food intake and time and energy investment.

Social behavior is typically intraspecific, and involves communication, which can be vocal or olfactory, involving pheromones. Pheromones are often involved in sexual signaling.

Dominance hierarchies suppress aggression by an "accepted" pecking order, and may be affected by hormones, sex or age. Territories are a subset of the home range, and involve defense of an area. Territories allow an animal to possess some limiting factors, such as a breeding site, mates, or food.

Sexual social behavior maximizes reproductive output. Most birds form a monogamous pair bond, which optimizes continuous incubation and then feeding of the offspring.

The social insects, particularly the Hymenopterans and the termites have a rigid, complex social structure, which involves a reproductive queen and nonreproductive worker animals. In the honeybees, the queen is the only reproductive animal, the females are nonreproductive workers, and the few males do not work and vie to mate with the queen. Because of the unusual haplo-diploid condition, it is genetically optimal for the workers <u>not</u> to reproduce, but to aid in the raising of their sisters. The workers are actually <u>more</u> related to their sisters than their own offspring, if they were to have any. The workers have a complex behavior associated with food finding, including the round and waggle dance.

Altruistic behavior is explainable when considering degrees of relatedness. For example, birds which help at the nest typically aid their parents in producing more offspring, which are their siblings, and therefore share many of the same genes. The study of sociobiology combines theories from population genetics, evolution and animal behavior to describe the adaptive significance of many social behaviors.

Lecture Outline

<u>Animal behavior</u>

- Responses of an organism to its environment
- Ethology is the study of behavior in the natural environment
- Biological rhythms are common; often innate
  - Circadian rhythms occur daily
    - Diurnal animals are active in the daytime
    - Nocturnal animals are active in the night
    - Crepuscular activities are active at dusk and dawn
  - Other behaviors follow the lunar cycle, particularly intertidal animals
  - Rhythms may be controlled by the pineal gland and hypothalamus
- Behaviors are genetically based
  - May be modified by the environment
  - Primarily influenced by the nervous and endocrine systems
  - May require physiological readiness to be expressed
  - Innate behaviors have a strong genetic component
    - Behavior referred to as fixed action patterns (FAP's)
    - Stimulated by a sign stimulus
  - May be modified by learning
    - Scope of learning varies among different species

<u>Conditioning</u>

- Classical conditioning
  - Connection made between normal body function and a stimulus
  - Pavlov and his dogs
  - Rang bell before feeding dogs; later salivated after hearing bells
- Operant conditioning
  - Positive reinforcement
    - Behavior to gain a reward
    - Animal experiments; rat presses bar, gains food pellet
    - A student might work for good grades for a scholarship
  - Negative reinforcement
    - Behavior to avoid a punishment
    - Animal experiments; rat presses bar to avoid electric shock
    - A student might work for good grades to avoid parental admonishment

Other simple behaviors

- Imprinting
  - Behavioral connection formed right after birth or hatching
  - Typically bond with parent
  - Studies by Konrad Lorenz
- Habituation
  - Ignore a repeated, meaningless stimulus
  - Wild animals become "used to" people

Insight Learning

- Involves application of past experiences to current situation
  - Most highly adapted in primates
- Learning is seen in members of nearly all phyla; abilities vary however

Behavioral ecology

- Migration
  - Regular, extensive movements of animals
  - Birds exhibit the greatest range of migration
  - Cues for migration include the sun, hormones, the earth's magnetic field, visual cues, and olfaction
    - Birds typically migrate by the stars
- Optimal foraging
  - Involves the most efficient way to gain food
  - Current topic of much research interest

Social behavior

- Animals are often found in aggregations
- Social behavior involves interactions between animals in aggregations
- Social behavior typically occurs between members of the same species
  - Social foraging involves cooperative hunting or feeding
- Societies are often found in social insects, fish, birds and mammals
  - Societies may involve cooperation, division of labor
- Requires communication between members of the society
  - Communication may be auditory or via scents
  - Pheromones are chemical signals
    - Particularly important in sexual signaling
- Dominance hierarchies
  - A means of regulating aggression in social species
  - Dominance may be affected by hormones, pheromones, sex of the individual
- Territories
  - Defended portion of the home range
  - Typically reflect a shortage of some resource, such as food, area or mates
  - Well studied in birds
    - Territories advertised by singing

Sexual behavior

- Sexual behavior is, by its nature, social
- Sex requires cooperation, suppression of aggression and communication
- Courtship rituals are often very complex
  - Males often display ritual aggression with other males
- Pair bonds assure cooperation in mating and raising of the young
  - Most birds are seasonally monogamous
    - Advantageous in incubating eggs, feeding young
    - Females typically invest more energy
- Play behavior of young develops adult behaviors

Eusociality

- Complex social behavior seen in bees, ants, wasps, and the unrelated termites
- Social behaviors are innate, rigid, and influenced by pheromones
- Honeybee social behavior
  - The queen is the reproductive female
    - Produce haploid eggs
  - Workers are nonreproductive females
    - Workers are diploid; a result of haploid sperm fertilizing haploid egg from the queen
  - Drones are non-working males; only function is fertilization of queen
    - Drones are haploid
    - Their sperm contain all of their chromosomes
  - Complex division of labor by the different ages of workers
  - Workers (sisters) are more closely related to each other than any potential offspring, so it is advantageous to work and take care of the queen's offspring (more of their sisters)
- Honeybee communication
  - Karl von Frisch showed that the dance of the workers shows the location of the food source
    - Round dance shows a nearby food source
    - "Waggle" dance shows a distant food source

Sociobiology

- Vertebrate systems exhibit complex and adaptive behaviors
- Altruism appears to benefit other members of the species, not the animal exhibiting the behavior
  - Based on relatedness between animals; perform altruistic acts on related animals, which share many of the same genes
  - Bird "nest helpers" typically help their parents raise another brood
- Sociobiology is a synthesis of genetics, evolution and ethology
  - E.O. Wilson
  - "The selfish gene"- behavior promotes an organism's genes being passed to the next generation

Research and Discussion Topics

• Birds are typically monogamous, on a seasonal basis. Other birds are polygynous (one male; more than one female mate), and a few are polyandrous (one female; more than one male mate). Give examples of birds which are polygynous and others which are polyandrous. What conditions might be conducive for the development of these alternative breeding systems?

Teaching Suggestions

• To differentiate between various types of behavior, I try to give real-life examples. For example, if a student is rewarded monetarily after receiving good grades, which type of behavior is that? A mother may ignore the demands of her young children, while you may find her children quite distracting. What type of behavior is the mother exhibiting?

• It is often difficult for students to understand the haplo-diploid system seen in the eusocial Hymenopterans. I typically put up this chart exhibiting the relatedness. The asterisk shows that workers are related by 3/4 to their sisters, which promotes their altruistic behavior.

Degrees of relatedness:

	Female bee (worker)	Male bee (drone)
Mother	1/2	1
Father	1	0
Sister	3/4***	1/2
Brother	1/4	1/2
Son	1/2	0
Daughter	1/2	1
Nephew	3/8	1/7

Suggested Readings

Diamond, J. "Reversals of fortune." *Discover*. April 1992. 70-75. Description of polygyny and polyandry in the animal world.

Cohen, J.P. "Naked mole rats." *Bioscience*. (42 (2): 86-89.
and
Sherman, P.W., J.U.M Jarvis and S. H. Braude. "Naked mole rats." *Scientific American*. August 1992. 72-78. Two articles on the the insect-like behavior of the eusocial naked mole rats.

Long, M.E. "Secrets of animal navigation." *National Geographic*. June 1991. 70-99.

Chapter 45. Ecology of Populations

Chapter Overview

The study of ecology includes the study of populations, communities and ecosystems. Ecologists may describe populations in terms of their density, as well as dispersion within its range. Due to interactions between individuals, members of the population may be uniform, random, or clumped in distribution.

Populations may be numerically described by the growth rate (r), which is a function of birth and death rates, emigration and immigration. The biotic potential is the maximum growth rate possible, and may be depicted as a J-shaped growth curve. However, due to limiting resources, populations typically exhibit an S-shaped growth curve, stabilizing near the carrying capacity of the environment.

Limits on population growth may be described as density dependent factors, which increase in effect as the population increases in size, and density independent factors, which are mostly abiotic factors like weather and natural disasters. Most density dependent factors involve limiting resources, or interspecific interactions like competition.

Organisms with a very high r are known as r-strategists, and are characterized by small size, asexual reproduction, and are often weedy or opportunistic species. K strategists are large, grow and reproduce slowly, and include many endangered animals.

The human growth so far has been characterized by a J-shaped curve, while approaching a so-far unknown carrying capacity. Demographers classify human populations into highly developed (low birth rates, high average GNP's), moderately developed, and less developed (high growth rates, low industrialization) countries.

Demographers also define four demographic sequential states, which the developed countries have already passed through. The preindustrial stage is characterized by high birth and death rates, but death rates drop in the transitional stage. In the industrial stage, the birth rate also drops, so population growth rates drop. In the postindustrial stage, the population is more affluent, and educated, and growth rates continue to drop. Age structures may indicate the future growth patterns of a country. If the proportion of prereproductive people is high, future growth may be anticipated. Because 1/3 of the human population is younger than 15, we will see continued growth in the future, particularly in the developing countries.

Lecture Outline

Ecology

- Study of organisms and their interactions between other organisms and the environment
- Population: members of the same species living in the same place at the same time
- Community: all of the populations in a given place and time
- Ecosystem: the communities and the physical environment in a certain area
- Biosphere: all of the ecosystems on Earth
- Ecosphere: the biosphere, the lithosphere, the atmosphere and the hydrosphere

Characteristics of populations

- Density
  - Number of individuals per unit area
- Range
  - The geographic limit of a population
  - Dispersion within the range
    - Uniform dispersion is even spacing
      - May be due to territoriality, competition, agonistic interactions
    - Random dispersion
      - Often due to lack of intraspecific interactions
      - Not commonly seen
    - Clumped dispersion
      - May be due to social interactions, asexual reproduction

Population growth

- Birth rate= b
  - Number of births per year (per 1000 people when describing humans)
- Death rate= d
  - Number of deaths per year
- Growth rate= r
  - $r = b - d$
- Doubling time
  - The amount of time for a population to double in size
  - $t_d = 0.7/r$
- Migration
  - Immigration is the addition of new members of the population (i)
  - Emigration is the loss of members of the population (e)
- Total growth rate, $r = (b - d) + (i - e)$
- Biotic potential
  - Maximum possible growth rate
  - Influenced by age at first reproduction, number of offspring possible per litter or clutch

- Larger organisms typically have smaller biotic potentials
    - Consider whales versus mice
- Exponential growth results from growing at the biotic potential
    - J-shaped growth curve

Limits on population growth

- Environmental resistance limits J shaped growth
- Carrying capacity (K) is the largest population that can be sustained for a continued period of time
    - S shaped (sigmoid) growth curve
    - Studies by Gause on *Paramecium*
    - When a population exceeds K, a crash may follow
- Density dependent mortality factors
    - Have increasing effects as population size increases
    - Includes predation, disease and competition
        - Classic studies on lynx and snowshoe hare
        - Includes interspecific and intraspecific competition
- Density independent mortality factors
    - Limiting factors which are not tied to population size
    - Includes severe weather like blizzards or hurricanes; or fires
    - Difficult to describe factors which are not tied to population size

Life history strategies

- r selected species
    - Small body size, large litter or clutch size
    - Live in unpredictable environments; are opportunists
    - Include many pest or weedy species
    - May employ sexual reproduction
- K selected species
    - Large body size, small litter or clutch size
    - Include many large endangered animals
    - Often care for young for an extended time

Human population growth

- Showed J shaped growth curve so far
- Population reached 1 billion in 1800
    - 5 billion in 1987
    - Projected to reach 6 billion in 1999
    - Current growth rate is 1.6%
- Malthus, a British economist, stated that human population growth was increasing faster than food supply
- Current growth due to recent decreases in the death rate due to better medicine and sanitation
- Unknown K for the human population; where will it level off? or crash?

Human demographics

- Developed countries
  - Lowest birth rates, low infant mortality
  - Long life expectancy
  - Lower replacement fertility
  - High average GNP
- Moderately developed countries
- Less developed countries
  - Highest birth rates, highest infant mortality
  - Short life expectancy
  - Low average GNP
  - Highest doubling times
  - Higher replacement fertility
- Total global fertility rate is 3.5; much above the average replacement fertility

Demographic stages

- Preindustrial stage
  - Birth and death rates are high
  - Population grows at a moderate rate
- Transitional stage
  - Death rate lowers, birth rate is still high
  - Population increases rapidly
- Industrial stage
  - Birth rate declines, population growth slows
- Postindustrial stage
  - Low birth and death rates
  - Population is characterized by affluence, education

Age structure of human populations

- Percent of people at different ages
  - Current human age structure shows that 1/3 of all people are under the age of 15 (prereproductive)
- Pyramidal age structure indicates rapid growth rate
  - Estimates show that because of this rapid growth seen in developing countries, by the year 2020, 85% of all people will live in these developing countries
- Stable population shows equal numbers of people in prereproductive and reproductive ages
- Shrinking populations show more people in postreproductive ages

Research and Discussion Topics

- One of the major concerns of biologists concerned with preserving tropical diversity is the continued growth of the populations of these tropical countries. Look up the growth rates of several tropical countries, and describe the potential effects of this growth on deforestation.

Teaching Suggestions

• I try to impress upon my students that the #1 global problem is population. Every environmental concern, from pollution to deforestation is tied to the simple fact that there are just too many of us. We discuss what it will be like in 25 years, considering recent projections on population growth. Sometimes students just memorize terms like doubling time without thinking what doubling means: it means a need for a doubling of food production, health care, roads, and perhaps doubling of pollution, trash etc.

Suggested Readings

May, R.M. "Ideas in ecology." *American Scientist.* May-June 1986. 256-267. Modern approaches in ecology, includes a great reference list.

Chapter 46. Communities of Organisms

Chapter Overview

Communities are complex associations of populations, which may be classified as either autotrophs (producers), heterotrophs (primary and secondary consumers), or decomposers. The trophic structure results in coevolution between interacting species. Predators and prey (whether plant or animal) have coevolved traits that allow predators to be more efficient, and the prey to avoid predation. Plants have mechanical barriers to predation, and some employ chemical defenses. Animals may flee, be camouflaged, have chemical defenses, or mimic harmful species.

Symbioses are also characterized by coevolution between the partners. Mutualistic relationships benefit both species, and are very common. Commensalism is a symbiotic relationship in which one partner benefits, and the other is not affected, and is harder to demonstrate. Parasites harm the host, but benefit the parasite. Most parasites do not cause death of the host, and some, in fact are not even pathogenic.

The niche describes the role of the organism in the environment and is the result of various limiting factors. Because of interspecific competitive exclusion, the observed niche is the realized niche. In the absence of competition, an organism would exhibit the fundamental niche.

Species diversity varies greatly; and is greatest in continental habitats with low environmental stress, and equatorial climates which are geologically stable. Succession describes the changes in a community over time. Primary succession begins on bare rock; secondary succession typically occurs after a fire or in abandoned fields. Both ultimately result in a climax community, which is a relatively stable state.

Lecture Outline

<u>Community ecology</u>

- Aggregation of all populations living in the same area and time
- Trophic levels
    - Producers = autotrophs
        - Photosynthetic
        - Provide base of food chain
    - Consumers= heterotrophs
        - Herbivores are primary consumers
        - Carnivores are secondary consumers
        - Omnivores eat plant and animal material
        - Detritivores eat dead organic matter
        - Decomposers break down organic material
            - Typically bacteria and fungi
            - Important in recycling

Interactions among community members

- Competition (discussed in chapter 45)
  - Interspecific competition
- Predation
  - Predator eats prey
    - Includes herbivory
  - Coevolution between predators and prey
  - Predators may ambush or pursue their prey
  - Plants have evolved spines and wax to resist herbivory
    - Others have noxious chemicals
      - Example: milkweeds
  - Animals also have adaptations to avoid predation
    - May hide from predators, including camouflage coloration
    - May have chemical defenses, often with warning coloration
    - Batesian mimicry
      - Harmless animals resemble a harmful species
    - Mullerian mimicry
      - All harmful species have similar coloration
- Symbiosis
  - Mutualism
    - Both partners benefit
    - Example: *Rhizobium* in root nodules of legumes
    - Zooxanthelle in corals
    - Mycorrhizae are found in about 80% of all plants
  - Commensalism
    - One organism benefits and the other is not affected
    - Example: epiphytes on tropical trees
  - Parasitism
    - One organism benefits and the other is harmed
    - Parasites rarely kill their host
    - Pathogenic parasites cause disease, and perhaps death
    - Other parasites cause no disease symptoms

The niche

- The niche is the ecological role of an organism
  - Includes its habitat, and all interspecific and intraspecific interactions
- The potential niche is the fundamental niche
- Interspecific competition results in the realized niche; due to competitive exclusion
  - The realized niche is what is typically observed in nature
  - Studies by Gause on *Paramecium*
- Limiting factors
  - Restrict the parameters of the niche

Species diversity

- Inversely related to geographic isolation
  - Islands are typically less diverse than continental communities

Inversely related to environmental stress
    Polluted areas have low diversity
    Polar regions are less diverse than tropics
Typically greatest at the margins of communities
    Edge effect (ecotones)
Introduced species often reduce the diversity of the community
Highest diversity in old, stable communities (like the tropics)

<u>Succession</u>

A progressive change in a community over time
Climax community is the "pinnacle" of succession
    Climax is not the static state that was once thought
Primary succession begins with a previously uninhabited area
    Pioneer community develops, often lichens
    Development on bare rock takes centuries to develop a climax community
    Primary succession on sand dunes is well studied
Secondary succession occurs after a fire, or on abandoned farmland
    Old field succession has been highly studied
        First plant species are weedy, later shrubs, then trees
        Animal successional stages also seen

Research and Discussion Topics

• How is diversity related to successional stages? Is diversity greatest at early, middle or climax stages?

Teaching Suggestions

• It is important to teach succession in a "modern" context. In a survey of subscribers to *Ecology*, in a list of the most "important" concepts in ecology, succession was #2, preceded only by the ecosystem. For so many years, it has been taught that succession is linear, and the climax is static. Instead, succession is a dynamic process, including plants <u>and</u> animals (as well as microorganisms) which leads to a dynamic climax stage.

Suggested Readings

Luoma, J.R. "Restless dunes." *Audubon*. Nov/December 1994. 78-89. Succession and conservation of the Great Lakes dune ecosystem.

McRae, M. "Why fight it?" *Audubon*. Nov-December 1994. 74-76, 121. The fire season of 1994 and the controversy over how we fight fires.

Watt, K.E.F. "Deep questions about shallow seas." *Natural History*. July 1987. 61-65. Explanations for the great diversity seen in shallow oceanic regions.

Boucher, D.H. "Growing back after hurricanes." *Bioscience*. March 1990. 163-166. Hurricanes and rain forests.

Chapter 47. Ecosystems

Chapter Overview

Biogeochemical cycles include carbon, nitrogen and water cycles (these have an atmospheric component) and the phosphorus cycle. Carbon cycles between the form of carbon dioxide and organic molecules, entering the biotic portion of the cycle through carbon fixation, specifically photosynthesis. It may be stored for long periods in trees, or even longer as fossil fuels or in oceanic sediments.

Bacteria are the key players in the nitrogen cycle. Bacteria in the soil or water fix atmospheric molecular nitrogen. Ammonia is then converted by nitrifying bacteria into nitrite and nitrate. These forms are then assimilated by plants. Ultimately, organic nitrogen compounds are metabolized and this produces ammonia. This ammonia may then be reassimilated, or lost to the atmosphere by the action of denitrifying bacteria.

Phosphorus becomes available to living things by erosion of phosphorus-containing rocks. It is then available to plant roots or phytoplankton, and it then passes through the food chain, and ultimately is broken down by decomposers. Like the carbon and nitrogen cycle, the phosphorus cycle has been greatly affected by the activities of humans.

Water cycles between oceans, the atmosphere, the land and organisms in the hydrologic cycle. It cycles slowly through groundwater and ice caps and glaciers.

In contrast to biogeochemical cycles, energy flows through the ecosphere in a linear manner. The flow of energy may be described by food chains or webs. Trophic levels can be used to illustrate pyramids of numbers, biomass and energy. Primary productivity measures the accumulation of plant matter; gross productivity is the total production, net productivity accounts for respiratory losses.

The ultimate source of energy is solar radiation, which provides energy for photosynthesis, and causes climatic patterns. Uneven heating of the earth's surface results in the warmer tropics and cold poles, and produces the movements of air and oceanic currents. These flows then influence the patterns of precipitation.

Lecture Outline

Ecosystems and cycling

Ecosystems are characterized by cycling of matter: biogeochemical cycles
Energy flows through the ecosystem; it does not cycle

Carbon cycle

All living things are composed of carbon-based molecules

Carbon dioxide is present in the atmosphere; makes up 0.03%
Also dissolved in water, tied up in rocks like limestone
Photosynthetic organisms fix carbon; incorporate it in sugars and other organic compounds
Organisms respire; typically by the process of aerobic respiration
Utilize sugar; give off gaseous CO_2
Carbon may be stored in wood of trees
May also be tied up in fossil fuels
We release carbon dioxide by burning fuels
Greatly increase amount of carbon dioxide released
Tied to global warming- "greenhouse effect"
Carbon is also tied up in the shells of marine organisms
May contribute to sedimentary rock

Nitrogen cycle

Nitrogen gas composes nearly 80% of the atmosphere
Molecular nitrogen is stable
Nitrogen fixation by bacteria converts N_2 to NH_3
Also accomplished by combustion, volcanoes, lightning and industrial processes
Use nitrogenase in the absence of oxygen
Terrestrial nitrogen fixers include *Rhizobium* in root nodules
Important aquatic nitrogen fixers are cyanobacteria, with specialized heterocysts
Nitrification also accomplished by nitrifying bacteria
Convert ammonia to nitrite then nitrate
Bacteria gain energy from these processes (are chemotrophs)
Assimilation
Plant roots absorb nitrate or ammonia
Use inorganic nitrogen to make organic molecules
Ammonification
During metabolism of proteins, produce ammonia
Animals may convert ammonia to urea or uric acid
These compounds may be excreted
Bacteria decompose organic matter, release ammonia
Ammonifying bacteria
Ammonia is now available to be recycled
Denitrification
Denitrifying bacteria convert nitrate to gaseous nitrogen
Are anaerobic
Represents a net loss of available nitrogen to the ecosystem
Human intervention in the nitrogen cycle
Addition of fertilizers, which may run off and stimulate algal blooms
Nitrates may contaminate groundwater

Phosphorus cycle

- No gaseous state; cycles between sediments and organisms
- Erosion of rocks makes phosphorus available to plants
  - Animals obtain phosphorus by eating
- Decomposers make phosphorus available to plants
- Cycles in aquatic habitats in similar way; except phosphorus is dissolved in water
- Phosphorus losses in marine deposits, erosion from land
- Human intervention in the phosphorus cycle
  - Phosphorus in agricultural products is transported great distances
  - Phosphorus fertilizer is used to replace loses

Hydrologic cycle

- Precipitation
  - Rain over land masses is from oceanic evaporation or transpiration of plants
- Rainwater may run into rivers and lakes, and ultimately reenter the ocean
  - Rainwater may enter the groundwater
    - Cycling of water in groundwater is slow
- Water may be held for long periods of time in icecaps and glaciers

Energy flow in ecosystems

- Sunlight is the energy input, used by photosynthesizers to form organic molecules which store energy
- Energy flow may be traced in simple food chains or complex food chains
  - Simple or complex; both are linear
  - Losses of energy between links shows why most food chains are relatively short
- Ecological pyramids
  - Pyramid of numbers shows the abundance of organisms at each trophic level
  - Pyramid of biomass shows the amount of living material at each trophic level
  - Pyramid of energy shows the amount of calories present in each trophic level
    - Cannot be inverted
- Productivity is a measure of energy transformations
  - Gross primary production is the total amount of photosynthesis
  - Net primary production includes the respiratory losses
  - Are rates; expressed as the amount of kilocalories/area/time, or the amount of carbon incorporated/area/time

Climate affects the distribution of living things

- Solar radiation
  - Some is reflected immediately
    - All is ultimately lost by radiation
  - Required for photosynthesis

- Warms the earth unevenly
  - Equatorial regions receive direct sun rays; higher temperatures
  - Polar regions have lower temperatures
- Seasons are a result of the tilting of the earth
  - Summers occur when that hemisphere is tilted toward the sun
- Atmospheric circulation is driven by unequal heating
  - Warmed air rises, moves toward the poles, cools
- Ocean currents are driven by the winds and rotation of the earth
  - Circular ocean currents (gyres)
- Precipitation patterns are based on water and air movements, and land forms
  - Tropical areas; air rises, moisture condenses, high rainfall
  - As moisture laden air moves over a continent, rises over mountain ranges
  - Deserts form in rain shadows

Research and Discussion Topics

• Investigate situations in which pyramids of numbers and pyramids of biomass are inverted. Give examples. Why can pyramids of energy never be inverted?

• Where might you find the least complex food webs? Where would the most complex food webs occur? In what situations are food chains extremely short? Where are the longest food chains found?

Teaching Suggestions

• My students find this handout comparing the cycles useful:

	Carbon	Nitrogen	Phosphorus	Water
Inorganic form?				
Atmospheric component?				
Bacterial Involvement?				
What form is taken up by plants?				
Long term storage in:				

• Point out the importance of bacteria in the nitrogen cycle. Bacteria truly "run" this cycle. Also differentiate between the "good guys" (the nitrifying and the ammonifying bacteria) and the "bad guys" (the denitrifying bacteria). In clear-cut forests and eroded lands, the denitrifyers dominate, which makes a bad situation worse, as available nitrogen is lost. This lessens the chance for revegetation.

Chapter 48. Major Ecosystems of the World

Chapter Overview

Terrestrial ecosystems are called biomes, which have similar assemblages of plants and animals because of similarities in climate and soil types. In the Northern Hemisphere, the treeless tundra is the northern-most ecosystem. It is characterized by organisms which concentrate their life cycles during the short summer. During the summer it is quite wet and supports huge populations of insects.

Adjacent to the tundra is the taiga, which does support trees and is highly exploited for lumber by humans. The temperate rain forests are also important producers of conifer and angiosperm trees. High precipitation, coastal fog and acidic soils characterize this biome. The temperate deciduous forest previously covered vast areas of North America, but large areas have now been replaced by agricultural lands.

Also converted to agriculture are the temperate grasslands in the US. The biome which previously supported tallgrass and shortgrass prairie has been replaced by monocultures of corn and wheat. The Mediterranean climate supports a chaparral biome, which is dominated by short, dense growths of drought-resistant evergreen shrubs. Deserts are even drier, and are dominated by small plants and animals which have striking adaptations to sparse precipitation. The African Savanna is characterized by high temperatures, variable precipitation, and supports grasses, and few trees. Large populations of hoofed mammals are found in the savanna.

Tropical rain forests are characterized by high precipitation, little annual variation in temperature, and extremely high diversity and productivity. The tall trees support several layers of communities. Tropical rain forests are particularly endangered due to human exploitation.

Biomes change with varying latitude, as well as altitude. Alpine tundra on mountaintops are similar to Arctic tundra, but receive more precipitation, and lack permafrost.

Aquatic ecosystems have different limiting factors; temperature and water are less important, but variations in salinity, current, and the tides affect distributions of organisms. In aquatic ecosystems, small organisms may be suspended in the water current (plankton), they may actively swim (nekton), or may dwell on the bottom (benthos).

Freshwater ecosystems include rivers and streams, which are influenced by current, and often pollution, and depend on external inputs of organic matter. Streams are characterized by zonation from the headwaters to the mouth. Lakes are characterized by zonation within the body of water; from near-shore to open water to aphotic benthos. Seasonal changes in temperature stratification are marked in temperate regions.

Estuaries are the interface between freshwater and marine ecosystems. Salt marshes and mangrove forests are extremely productive ecosystems. Marine environments may be divided into the intertidal, the pelagic (neretic and oceanic provinces) and the benthic zones.

Lecture Outline

Biomes
- Distinct terrestrial ecosystem
- Share similar climates, soil, plants and animals

Tundra
- Extreme northern latitudes where snow melts annually
- No land masses in the Southern Hemisphere at appropriate latitudes
- Short growing season, but long days
- Little precipitation
- Soils are nutrient-poor, low in organic material
  - Soil underlain by permafrost, which prevents drainage
  - Bogs and shallow lakes
- Diversity is low, numbers of animals may be very high, particularly insects
  - No reptiles or amphibians
- Humans interfere in oil exploration, military use

Taiga, or boreal forest
- Also primarily in Northern Hemisphere
- Little precipitation, soils are nutrient poor, acidic
- Permafrost, when present, is deep
- Many bodies of water
- Deciduous trees, conifers
- Most animals are small, some large mammals
  - Abundant insects, few reptiles and amphibians
- Humans interfere in trapping mammals, harvesting forests

Temperate rain forests
- Northwest coast of Northwest US, also parts of Australia, S. America
- High precipitation, dense coastal fogs
- Soil is nutrient-poor
- Large evergreen trees, epiphytes
- Important source of lumber; often replaced with monoculture of trees

Temperate deciduous forest
- Great seasonal variation in temperature
- Moderate precipitation
- Rich soils

Previously contained many large mammals
Many reptiles and amphibians, insects
Great influence by humans; cut forests, planted farms

Temperate grasslands
Great seasonal variation both in temperature, precipitation
Many of the grasses are sod-formers
Few trees, except in riparian areas
Tallgrass prairies in areas of higher precipitation; shortgrass prairie in dryer areas
Little prairie left; much has been farmed
Important areas for growing corn and wheat (which are also grasses)

Chaparral
Mediterranean climates; found around Mediterranean Sea
Also found in California, Australia, Chile, S. Africa
Dense growth of drought resistant low shrubs
Soil is thin, low in organic material
Fires often occur in late summer and fall
In California, these fires often involve loss of homes, and subsequent mudslides in the following winter

Deserts
Found in both temperate and tropical areas
Marked diel temperature fluctuations
Plants are often allelopathic, resulting in uniform distribution
Animals tend to be small, many water conserving adaptations
Human influence involves off-road vehicles, developments which need much irrigation

Savanna
A tropical grassland, has few trees, typically *Acacia*
Little seasonal change in temperature, but much variation in seasonal precipitation
Soil is relatively infertile
Plants exhibit adaptations for low precipitation, fires
Supports great herds of hoofed mammals
Human influence includes conversion to rangeland for cattle

Tropical rain forests
High precipitation, temperatures
Soils are old, nutrient poor; most nutrients are tied up in biomass
High diversity, productivity
Trees are evergreen angiosperms
Layered ecosystem
Crowns of tallest trees

Middle story
- Many epiphytes
- Lianas are used by animals to travel from tree to tree, and up and down

Understory; adapted for life in little light

Human influence has been devastating in the last few decades

Altitude effects on ecosystems

- See changes in ecosystems as go up in altitude
- Tundra on mountaintops is called alpine tundra
  - Tundra in the north is called arctic tundra
  - Alpine tundra lacks permafrost, has relatively high precipitation

Aquatic ecosystems

- Characteristics
  - Temperature is moderated due to high specific heat of water
  - Water is not a limiting factor
  - Salinities are variable; freshwater is hypotonic to vertebrates, salt water is hypertonic to vertebrates
  - Light does not penetrate to the bottom of all lakes, or most oceanic areas
- Aquatic organisms
  - Planktonic organisms are at the mercy of the currents
    - Phytoplankton are photosynthetic; the producers
    - Zooplankton are protozoans, small animals, and larvae
  - Nekton are able to swim against the currents
  - Benthic organisms live on the bottom; are attached, burrow, or are mobile

Freshwater ecosystems

- Small total area; but very diverse habitat and many species
  - Important in the hydrologic cycle
- Rivers and streams
  - Variable habitats from source to mouth
  - Current affects organisms
  - Depend on watershed for inputs
  - Human impacts include water pollution, dams and channelization
- Lakes and ponds
  - Littoral (near-shore) zone
    - Emergent angiosperms
  - Limnetic zone
    - Open water, photic zone
  - Profundal zone
    - No light; no photosynthesis
  - Stratification
    - In summer, warmer layers on top, colder at bottom, separated by a thermocline

In autumn, fall turnover
Temperatures are the same from top to bottom
In winter, stratified, with ice at the top and 4° C water below
In spring, another turnover
Spring turnover stimulates algal growth
Freshwater wetlands
Includes bottomland forests, prairie potholes and bogs
Highly productive, important ecosystems

<u>Estuaries</u>
The interface between fresh and salt water
Salt marshes are dominated by grasses
Are extremely productive due to:
Tidal circulation
Nutrients from rivers
Productive plants
Many fish reproduce in estuaries
Most estuaries have been severely degraded or dredged
Mangrove forests are found in tropical areas

<u>Marine ecosystems</u>
Influenced by tides, waves and currents
Intertidal zone
Very diverse, productive
Organisms must be able to deal with tides and waves
Pelagic zone
Open ocean
Euphotic region
The lit region; down to 100 m
Neritic province- down to 200 m
Covers most of the continental shelf
Oceanic province- deeper than 200 m
High pressure, cold, no light, scarcity of food
Animals are scavengers or predators

<u>Interaction between ecosystems</u>
Ecotones are the boundaries between ecosystems
Many animals migrate between ecosystems

Research and Discussion Topics

• Which biomes or oceanic zones are characterized by the highest diversity? Lowest diversity? Highest and lowest biomass? What animal groups are restricted in their distribution in biomes and zones?

Teaching Suggestions

• A chart for students to fill out may help them compare and contrast the characteristics of each of the biomes:

Biome ______________

Where found:

Precipitation
Relative amount
Seasonality

Temperature
Relative temperature
Seasonality

Soil type

Major plant associations

Major animal associations

Human influences

Suggested Readings

Glantz, J.M.H. "Drought in Africa." *Scientific American* June 1987. 34-40. Climate and politics interact to produce drought conditions.

Chapter 49. Environmental Problems

Chapter Overview

Extinctions are currently occurring at a rate never experienced on earth. The extinction rate is much higher than the naturally occurring background rate, and is attributable to human activities: destruction of habitats, pollution, introduction of exotic species, and purposeful killing for commercial, sport or sustenance reasons.

Conservation efforts involve preservation in situ in parks and reserves, and ex situ captive breeding efforts; typically aimed at the most severely endangered species. Plants are endangered as well; deforestation of tropical and temperate forests is occurring at an alarming rate. Tropical forests are being logged for slash and burn agriculture and fuelwood gathering. These are primarily problems associated with overpopulation in developing countries. Also, the consumption of timber and beef by the developed countries is resulting in tree loss by commercial logging and clearing for cattle ranching.

A variety of chemicals contribute to the "greenhouse gases", which may result in global warming. The predicted increases in carbon dioxide and CFC's, for example, may result in a global increase in temperature of 2-5° C, which will cause sea level to rise, patterns of precipitation to change, and the effects on plants and animals may be very significant. Response to global warming involves prevention, mitigation and adaptation. All of these environmental problems are interrelated, with the ultimate cause being human overpopulation.

Lecture Outline

<u>Extinction</u>

- A natural process, but greatly accelerated by humans
- 1/4 of the higher plant families may become extinct by end of next century
- Endangered species are in eminent danger of becoming extinct
  - Threatened species also have decreased numbers

<u>Causes of endangerment or extinction</u>

- Destruction of habitat
  - Cities, agriculture, logging
- Pollution
  - Acid, chemical, thermal pollutants
- Introduction of foreign species
  - Exotic species have fewer controls on their population
- Deliberate control measures
  - Seen in species viewed as pests, like prairie dogs
- Commercial hunting
  - May be legal or illegal, like hunting of rhinos and bears
- Sport hunting

        Relatively well controlled in most countries
    Subsistence hunting
        Particularly important in the past
    Commercial harvesting for zoos, research and the pet trade

Conservation efforts
    In situ conservation includes parks and reserves
        In the US, 3000 national parks and protected areas
        Many have multiple uses, including hunting and mining
    Ex situ conservation includes captive breeding programs, seed banks
        Artificial insemination
        Embryo transfer (host mothering)

Deforestation
    Removal of trees accelerates erosions, desertification, flooding problems
        Also affects global precipitation patterns
    Problems in the tropics:
        Subsistence agriculture = slash-and-burn agriculture
            Accounts for the majority of deforestation in the tropics
            Due to overpopulation in developing countries
        Commercial logging
            Occurs much more rapidly than is sustainable
        Cattle ranching
            Primarily in South and Central America
            Many owned by large multinational corporations
            Much of this beef goes to the processed food or fast-food industry
        Wood cutting for heating and cooking

Global warming
    "Greenhouse gases" include carbon dioxide, methane, ozone, nitrous oxide and chlorofluorocarbons
        Atmospheric carbon dioxide has increased dramatically since the Industrial Revolution
        Burning in the tropics increases carbon dioxide levels
        CFC's come from old refrigerators and air conditioners
        Much methane comes from landfills
    Doubling of carbon dioxide may result to a 2 - 5° C increase in global temperature
        May occur within the next 50 years
        Will cause sea level to rise as much as 2.2 meters; import in low-lying countries like Bangladesh
        Will change precipitation patterns, and probably agricultural production
        Most sensitive ecosystems may be arctic oceans, coral reefs, mountains, coastal wetlands, tundras, and northern forests
        Greatest impact on plants because they are stationary

Approaches to global warming:

Prevention

Alternatives to use of fossil fuels, such as solar energy

Mitigation

Plant trees, increase efficiency of automobiles

Adaptation

Move inland!

Develop new crops which are heat-resistant

<u>Ozone depletion</u>

Ozone, O_3, is a naturally protective molecule which screens out UV radiation

Depletion of the ozone layer over Antarctica

CFC's are implicated; a single chlorine molecule can destroy thousands of ozone molecules

CFC's come from refrigerants, cleaners, pesticides, and other manufacturing processes

Increases in UV light are linked to development of cataracts, skin cancer, suppressed immune systems

May also damage plankton, agricultural crops

Even though CFC's are being phased out, the molecules are very stable and will continue damage for many years

<u>Other environmental problems</u>

Solid waste

Are filling up

Contribute methane to atmosphere

Refrigerators leak CFC's

Overpopulation

Obvious as a key to most environmental problems

Research and Discussion Topics

• If global warming continues at the pace suggested by some scientists, we may need to develop agricultural crops which can withstand increasing temperatures. Remembering the photosynthetic adaptations discussed in previous chapters, what types of plants (which photosynthetic pathways) should be considered?

Teaching Suggestions

• I've always believed that this chapter should be taught at the beginning of the semester, even though we typically save it for the end. Everyone hears of the ecological buzzwords like ozone hole and greenhouse gases, but few students really understand how it all works. This is our chance to make our students more ecologically aware!

Suggested Readings

Moore, R. "Changing our attitude about the environment." *American Biology Teacher*. March 1992. 54 (3): 132-133. A short article discussing environmental problems associated with overpopulation.

Cohen, J.E. "How many people can earth hold?" *Discover*. November 1992. 114-119. Population growth estimates and the accompanying environmental impacts.

Repetto, R. "Deforestation in the tropics." *Scientific American*. April 1990. 36-42. Logging and clearing forests are devastating tropical rain forests.

Thompson, J. "East Europe's dark dawn." *National Geographic*. June 1991. 36-67. The terrible air and water pollution in Eastern Europe, and their effects on humans.